GW01315910

MEASUREMENT OF GRASSLAND VEGETATION AND ANIMAL PRODUCTION

edited by

L. 't Mannetje

CSIRO Division of Tropical Crops and Pastures

Cunningham Laboratory

Brisbane, Queensland, Australia

BULLETIN 52

Commonwealth Bureau of Pastures and Field Crops,

Hurley, Berkshire, England

COMMONWEALTH AGRICULTURAL BUREAUX

Published in 1978 by the Commonwealth Agricultural Bureaux
Farnham Royal, Bucks, England

ISBN 0–85198–404–5

Printed in Great Britain by The Cambrian News (Aberystwyth) Ltd.,
Aberystwyth

LIST OF PLATES

LIST OF CONTRIBUTORS

Dr. J. L. Corbett. Senior Principal Research Scientist with the Commonwealth Scientific and Industrial Research Organization (CSIRO) Division of Animal Production at Armidale, New South Wales, Australia. From 1944 till 1947 he attended the University of Reading, England and after receiving the degree of BSc (Agric.) proceeded to Massey Agricultural College, University of New Zealand, where he was awarded the degree of MAgrSci and DSc. In 1950 he returned to the United Kingdom, where he was attached to the Rowett Research Institute, Aberdeen. In 1963 he took up his present appointment.

Dr. M. B. Dale. Principal Research Scientist, CSIRO Division of Tropical Crops and Pastures (formerly Tropical Agronomy), Brisbane, Queensland. Received his BSc and PhD degrees from the University of Southampton, England, where he studied Botany under Professor W. T. Williams from 1957 to 1964. He had short-term research and lecturing engagements at the Universities of Sheffield and Hull, England, and of Michigan, Ann Arbor, USA. In 1968 he joined the CSIRO Division of Plant Industry, Canberra, but returned to England in 1970, where he was involved in systems ecology at the Grassland Research Institute. He has been in his present position since 1972.

Mr. A. D. Johnson. In charge of analytical chemistry at the CSIRO Cunningham Laboratory, Division of Tropical Crops and Pastures, Brisbane, Queensland. He graduated as BSc in 1957 and was awarded MSc in 1976 from the University of Queensland. Before taking up his present employment in 1970 he was with the Queensland Department of Primary Industries and the CSIRO Tobacco Research Institute at Mareeba, Queensland.

Dr. L. 't Mannetje. Principal Research Scientist, CSIRO Division of Tropical Crops and Pastures, Brisbane, Queensland. Born in the Netherlands, where he studied at the Landbouwhogeschool, Wageningen, from 1951 to 1958. Specialising in grassland ecology, he was awarded the degree of Landbouwkundig

Ingenieur. He took up his present position in 1958, but was seconded to the Malaysian Agricultural Research and Development Institute, Serdang, Malaysia, in 1974 and 1975. He received the degree of PhD from the University of Queensland in 1967.

Dr. F. H. W. Morley. Chief Research Scientist with the CSIRO Division of Plant Industry, Canberra. Received the Hawkesbury Diploma in Agriculture in 1937 and graduated as Bachelor of Veterinary Science at Sydney University in 1942. From 1943 to 1954 he was attached to the New South Wales Department of Agriculture. In 1950 he was awarded the degree of PhD from Iowa State University, USA and in 1954 he joined CSIRO. A Fellow of the Australian College of Veterinary Science and of the Australian Society of Animal Production he was awarded the Medal of the Australian Institute of Agricultural Science in 1970. In 1977 he joined the University of Melbourne.

Mr. G. A. McIntyre. Senior Principal Research Scientist with the CSIRO Division of Mathematics and Statistics in Canberra before his retirement in 1971. He was educated in Western Australia and obtained a BSc (Hons) degree as well as a Diploma of Education. He worked in education before joining the forerunner of CSIRO in 1940. As well as advising grassland and other research workers all over Australia on statistical design and analysis, he lectured in statistics at the Canberra Forestry School. After retirement he continued to work as a part-time consultant with the CSIRO Division of Wildlife Research, Canberra. He also maintained an active interest in updating statistical analysis programs, commonly referred to as the McIntyre programs, for the CSIRO computer network. He died in January 1974.

Dr. J. C. Tothill. Principal Research Scientist with the CSIRO Division of Tropical Crops and Pastures, Brisbane, Queensland. Born in New Zealand, he received his BAgrSci degree from Lincoln College, University of New Zealand in 1954. He then worked for the New Zealand Department of Agriculture in Central Otago before going to the University of California, Davis, in 1957 where he obtained the degree of PhD. He came to his present position in 1962.

FOREWORD

For nearly 200 years, attempts have been made to analyse natural vegetation and classify it into broad ecological groups. In the last century much synecological work was centred in academic institutions of Europe, but some botanists, as in Switzerland for instance, devoted time to detailed investigations of those aspects which concerned grassland vegetation in order to provide a necessary basis for assessing the quality and yield of grasslands, the ultimate goal being optimal animal production. An early example of studies of this kind in England can be seen in the handsomely produced reports of G. Sinclair (1816 and 1826) in which he wrote up the results of investigations on the Duke of Bedford's estate at Woburn.

In the course of time there has been an ever increasing demand for precision in such studies. How great are the demands now laid both on the grazing animal and on the research worker will be very evident in the chapters that follow.

All who were adult in the thirties of this century will recall the dust-bowl of the western and mid-western states of the USA, brought about by a combination of overgrazing, overcropping and a succession of dry years. In its endeavours to find a remedy for this catastrophe, the USA became both trial ground and growth centre for new techniques, especially those developed for sparse vegetation; and so, out of disaster, came new ideas and approaches. Another aspect that came to the fore at this time was a growing interest in the biotic factor as represented by the grazing animal. Indeed, it was a request from New Zealand for a collection of abstracts on this topic and the publication at about the same time of a report on the subject from Saskatchewan, Canada, that in 1943 gave the initial impetus to set about the task of compiling Bulletin 42, the forerunner of the present book. This Bulletin essentially summarized the state of the art up to the early 1950's and, with hindsight, this now seems to have been an excellent stop date. Since that date, there has been another great leap forward in pasture studies aided both by the development of entirely new techniques and instruments and by greater understanding of the biological system involved at all levels. The vastly increased quantity of information made available by the use of these instruments would have remained to a greater or lesser extent impossible to exploit fully but for another development, that of the computer for data handling and analysis. The computer has also brought back into prominence the earlier, simpler presence/absence methods, data from which can be handled so effectively by the binary system.

The present book differs from its predecessor in its approach to the subject, and it is the omission of the word 'methods' from the title that gives the key to the difference. While the earlier book, a textbook of methodology, sorts out and classifies from the then confusion the various concepts, criteria of measurement, methods, techniques and adaptations, the present book applies methods more directly to pastures within the framework of the farm and the economic setting and discusses what results are to be expected as conditions vary. It is for the reader who already has a fair understanding of the subject and of the new techniques available and who wants to know how best and where best to use them.

The question of publishing a second edition of Bulletin 42 dates back to 1965 when Professor C. M. Donald, then of the Waite Agricultural Research Institute of South Australia, lent his considerable authority and experience to the decision to do so. In the event, it proved very difficult to find an author or authors with both the professional competence needed and the equally necessary willingness to give of their time to the task. We have been fortunate in finding seven practising specialists, each able to write with authority on his chosen subject, to complete this work. Because of the many differences in approach between this book and its predecessor, it was decided to issue it not as a second edition of its forerunner, but as an entirely separate publication, a view with which I entirely concur.

The present book well illustrates the cross-fertilization of ideas contributed from all over the world in the post-1950 period, a process greatly enhanced by the overseas travel which has enabled Australian workers in particular to exchange views and experiences with their fellow workers in so many countries.

The grassland world will be grateful to the authors of this book, and especially to Dr. L. 't Mannetje who has taken on the responsibility of co-ordinating and editing the whole work, as well as contributing two of its chapters. In grateful acknowledgement of the co-operation and encouragement received from Australia and many other countries during the preparation of Bulletin 42, I am most happy to send its successor, Bulletin 52, on its way and wish it long and useful life.

DOROTHY BROWN
Henley-on-Thames

February 1977

PREFACE

This book is the joint effort of seven research workers, each writing about a field of grassland research in which he has a specialist interest. Its aim is to present a critical review of methods for measuring grassland vegetation in terms of botanical composition, quantity, chemical composition and animal production. In addition, there is a chapter on statistical aspects of sampling and one on computer methods of pattern analysis.

Dorothy Brown's book 'Methods of Surveying and Measuring Vegetation', which was published as Bulletin 42 of the Commonwealth Bureau of Pastures and Field Crops in 1954, was the first comprehensive treatise on this area of research methodology. It has always been in great demand, but during the 25 years since its publication there have been many new developments in grassland research, examples among which can be mentioned *in vitro* digestibility, capacitance meters, fistulation of animals and the use of computers. Another reason for a new handbook was the 'tropical pasture revolution' which has occurred during the last 20 years, and the need to take account of the differences in approach to grassland research that have been developed and which have been set out in detail in the previous publication in this series, Bulletin 51 'Tropical Pasture Research—Principles and Methods', edited by N. H. Shaw and the late W. W. Bryan and published in 1976.

The present volume differs from Dorothy Brown's Bulletin 42 in several respects. Firstly, it deals exclusively with vegetation used for herbivores, particularly domestic animals. Hence the word 'grassland' in the title. However, many of the methods described are equally applicable to non-grazed vegetation. Secondly, each chapter has been written by an authority in the field covered. Although this has inevitably resulted in differences in approach and style among authors, it has made it possible to bring together a range of expertise much beyond the reach of any author endeavouring to deal single-handed with the whole field of grassland research methodology. Each chapter is intended to be a self-contained unit; this may have led to some duplication of the topics covered, but because many readers are likely to read only those chapters that deal with their particular interests, I have not discouraged this.

This is not a recipe book, but rather is meant to be an introduction to grassland research methodology and to be a guide to the available techniques. Each contributor was asked to adopt a critical approach to the choice of methods presented and in his discussion of their applicability. In most cases it will remain necessary for investigators to refer to the original publication in order to be able to use most of the methods dealt with; for this reason, considerable importance has been attached to the provision of appropriate literature references.

Although all authors at the time of writing were working in Australia, they collectively represent a very wide range of experience and background. The subject matter and literature reviewed is in no sense restricted to any one part of the world, whether in the geographical or climatic sense. One of the contributors, Mr. G. A. McIntyre, died soon after he had prepared an advanced draft of his chapter, and Mr. R. L. Sandland very kindly undertook the task of completing it.

I wish to thank Dr. E. M. Hutton, then Chief of the CSIRO Division of Tropical Agronomy (now Tropical Crops and Pastures) for permission to undertake the task of editing this book, and Mr. T. W. Elich and Mr. K. H. L. van Bennekom for checking the literature references. Finally, I would like to thank Miss Dorothy Brown for having done me the great honour of writing a foreword.

L. 't Mannetje

July 1976 *Brisbane*

CONTENTS

Chapter 1

AN INTRODUCTION TO GRASSLAND VEGETATION AND ITS MEASUREMENT

L. 't Mannetje

It is essential to define grassland vegetation before methods for its measurement can be discussed. The word grassland implies land covered by grasses but grassland is most commonly defined as a plant community in which grasses are dominant, shrubs are rare and trees absent (e.g. Moore, 1964; Milner & Hughes, 1968; Spedding, 1971), thus allowing other herbs and even shrubs to be part of the vegetation, but not trees.

However, for the purpose of this book this definition, based entirely on vegetational criteria, is too narrow. It is recognised that the main function of grassland is to provide herbaceous feed for large herbivores, both wild and domesticated. Plant communities other than those covered by the common definition provide a substantial proportion of herbaceous animal feed. These are not necessarily grass-dominant, nor are they neccessarily free of trees. Take for example heaths and open woodlands (Plate 1). In this connection, the term grazing lands is often used (e.g. Brown, 1954; Moore, 1970; Whyte, 1974), but this in turn is too narrow, as it excludes grasslands that are never grazed. A satisfactory new term to denote this type of land-use is not easily found and I prefer to retain the term grassland, but to define it as "an ecosystem from which

PLATE 1. Open woodland with understorey of native grasses and *Stylosanthes humilis* in north Queensland, Australia

herbaceous animal feed is derived". The word ecosystem indicates that we are not only dealing with the vegetation as if it were an independent entity, but with the whole system of soil and vegetation as determined by climatic and edaphic conditions. This definition also links grassland inseparably with the animals that feed on it or from it.

This definition of grassland excludes lawns, parks and playing fields, which strictly speaking ought to be included in the more common definition. On the other hand, it includes heaths, tundras, open woodlands and green crops, such as oats grown for grazing, so long as these are used for the feeding of herbivores.

Grassland vegetation, therefore, may consist of grasses, legumes, other herbs, trees, shrubs and lower plants. Although not contributing to the diet of herbivores, inedible plants are important in that they affect the edible components of the vegetation through competition.

Ecologically, natural grasslands are a vegetation type controlled by a combination of climate, soil, topography, biotic factors and fire (Moore, 1964). However, many such areas are no longer under grassland vegetation because they are used for crop production, e.g. prairies in North America, pampas in Argentina, steppes in the USSR and downs in Australia. On the other hand, many areas previously under forest have been cleared and either native grasses and herbs or sown pasture species are being maintained by such activities as grazing, cutting, burning and fertilizing.

Thus it appears valid to divide grasslands into *natural* and *induced*, the latter being sometimes referred to as artificial or tame pastures. However, if it is accepted that induced pastures exist only by the virtue of interference from man, what should we call natural grasslands that have been modified by agricultural practices such as oversowing, fertilization, etc ?

Grasslands are often further subdivided into permanent or temporary (leys), or according to the structure of the vegetation into open and dense, tall, medium and short. This structure depends on climate and soil, but agricultural management practices also influence it. For instance, open grasslands—whether natural or induced—are usually associated with dry environments in which plant density is limited by the amount of available moisture. However, this does not mean that dry environments must always carry open grasslands, because management may be able to provide irrigation and thus achieve a dense grassland. A striking example of how management may change a vegetation type is the effect of grazing on the tropical legume *Leucaena leucocephala* (Plate 2).

PLATE 2. The browse shrub *Leucaena leucocephala*

In the ungrazed condition, it may grow as spaced trees up to 7 m in height or may form inpenetrable thickets as in parts of Oceania, but it can also be kept as a low growing shrub by strategic grazing, or it may behave essentially as a herb under continuous, heavy grazing. Another example of the effect of management on vegetation type is the choice of species mixtures when sowing grassland. In tropical and subtropical regions with good moisture supply one may either decide to have shrub-grassland (*Leucaena leucocephala*), or open tall-grass (e.g. *Pennisetum purpureum*), or open medium-height grass (e.g. *Setaria anceps, Panicum maximum*) or dense short-grass (e.g. *Digitaria decumbens, Cynodon dactylon*). These vegetation types may grow side by side in the same locality.

Other examples could be cited, but it suffices to state that climatic, edaphic and management factors may interact, so that various combinations result in different vegetation types in the same locality, or in similar vegetation types in different localities. It is therefore not very profitable to divide the world's grasslands into vegetation types.

Nevertheless, for the purpose of measurement there is some advantage in having certain standards for the description of the various forms of grasslands. For this purpose, Brown (1954) described a number of grassland types on the basis of ground cover (density), height and composition. She recognized *open grasslands, dense grasslands* and various *herb/grass/shrub* associations. The open grasslands she considered mainly as natural climax vegetation in regions which experience a distinct dry season and she subdivided these again on the basis of height (high, medium, short). Dense grasslands have less bare ground, but the Mediterranean-type annual grasslands (in regions with wet, cool winters and hot dry summers) consist largely of annuals which leave virtually bare ground in summer. In contrast, the dense perennial grasslands of the humid temperate regions, which have replaced forest climax vegetation, have a permanent cover with no bare ground. Within the various shrub associations, Brown recognized shrub/open grassland and shrub/dense grassland and dense shrub.

There is no need to change this classification for the purpose of measurements, except that we might add *tropical twining-legume/grass pastures* (Plate 3), the measurement of which poses special problems. Bulletin 51 of the Commonwealth Bureau of Pastures and Field Crops (Shaw & Bryan, 1976) gives a detailed account of the principles and methods of research into the development and evaluation of grassland in the tropics.

With regard to measuring grassland vegetation, the same principles apply to all types, but the nature, size, shape and number of sampling units may have to differ in order to take account of differences in density or growth habits. From the sampling point of view it is not important whether a grassland is natural or induced, permanent or temporary. It is, however, important to know whether the vegetation is a monoculture or a mixture, because methods for measuring mixtures are more complex.

PLATE 3. Tropical twining-legume grass pasture of *Desmodium intortum* cv. Greenleaf and *Setaria anceps*

Vegetation is measured for three main purposes. Firstly, to describe the status of the vegetation in terms of its floristics, ground cover, the amount of dry matter available, growth, the amount of nutrients available and so on. Secondly, to assess the changes brought about by management practices, either in experiments or on a farm scale, e.g. fertilizer use, grazing pressure, or grazing method. Thirdly, to determine the ability of the vegetation to provide feed for herbivores. In the latter context, the value of grassland is determined in terms of its ability to provide energy, protein, minerals and vitamins and also by the absence of toxins and deleterious hormones. Although these attributes can be measured directly, their levels in the vegetation do not necessarily indicate the level of animal production that can be obtained. Herein lies the great difficulty of measuring the productivity of grazed grassland. The ultimate product is the consumable animal commodity, and measurements of the vegetation itself are not always meaningful. The best measurement of productivity is that made in terms of actual animal performance. However, this is expensive and often not possible owing to the lack of area or lack of animals (many grassland research stations have no animals available for research). In such cases, the experimenter may have to be content with measuring intermediate stages in the production process, but he must always be aware that it is the consumption and conversion by the animal which finally determines the value of the vegetation. At the same time, this does not necessarily mean that all animal measurements are by definition meaningful. Animal performance in controlled conditions in pens, for example, is not equivalent to that of grazing animals. In the first place, the herbage is cut and fed to the animal away from the pasture, thus reducing the degree of herbage selection, and secondly the animal expends less energy in obtaining its diet in pens than while grazing in the open. Even measurements made on grazing animals can have an artificial element, for instance in the case of fistulated or tethered animals, or when a different type of animal is used in measurements as compared with that used under actual production conditions. Methods for measuring a plant community will differ according to the use that is to be made of the measurements. For example, the contribution of a species to the botanical composition of a community can be estimated in terms of frequency, basal cover or dry weight proportion (Chapter 3). From an animal nutrition point of view, the most important of these measurements would be the dry weight percentage, whereas long-term changes in botanical composition are better measured as frequency percentage, because this is less sensitive to short-term changes. Also, where herbage deterioration is a problem, total dry matter yield is not as important for animal nutrition as is amount of edible material.

In keeping with the definition of grassland as the supplier of feed for herbivores, the measurements of greatest interest are those concerning the plant community rather than the individual plant. The botanical composition of the community is important because it gives information on the elements of the community and because different plant species, varieties or cultivars differ with regard to climatic and edaphic requirements, have different growth rates, different chemical composition and different feeding values. Grassland yield can be expressed as dry matter produced over a certain period or available at a certain time, but to be really useful it needs to be further qualified; it may be expressed in terms of digestible nutrients, starch equivalent, green matter, crude protein yield and so on. The chemical composition and digestibility are not only necessary in order to calculate some of the expressions of yield mentioned above, but are important in their own right as measures of sufficiency or deficiency in an animal's diet.

Although the definition of grassland given at the beginning of this introduction excludes parks, sports fields and lawns, many of the methods of measuring vegetation attributes dealt with in this book are directly applicable to non-agricultural grassland. For example, the density and species composition of the turf on sports fields and lawns can be measured in the same way as on closely grazed dense pastures.

Grassland vegetation, by its very nature, is well adapted to the prevention of soil erosion and this aspect is as important on agricultural as on non-agricultural land. In both instances, well managed grassland vegetation offers stability to the environment, thus protecting one of the basic resources of the earth for the continuation of terrestrial life of all kinds, including that of man. Therefore, the measurement of attributes of an apparently non-agricultural nature may be as important as the measurement of those having immediate agricultural relevance.

REFERENCES

BROWN, Dorothy (1954) Methods of surveying and measuring vegetation. *Commonw. Bur. Pastures Field Crops, Hurley. Bull.* 42.

MILNER, C. & HUGHES, R. E. (1968) Methods for the measurement of the primary production of grassland. International Biological Program. Handbook 6, Oxford : Blackwell Scientific Publications.

MOORE, C. W. E. (1964) Distribution of grasslands. *In* Grasses and grasslands. Ed. C. Barnard, Melbourne: Macmillan & Co. Ltd., pp. 182-205.

MOORE, R. M. (Ed.) (1970) Australian grasslands. Canberra: Australian National University Press.

SHAW, N. H. & BRYAN, W. W. (Eds.) (1976) Tropical pasture research. Principles and methods. *Commonw. Bur. Pastures Field Crops, Hurley, Bull.* 51.

SPEDDING, C. R. W. (1971) Grassland ecology. Oxford: The Clarendon Press.

WHYTE, R. O. (1974) Tropical grazing lands : communities and constituent species. The Hague: Dr. W. Junk, NV.

Chapter 2

STATISTICAL ASPECTS OF VEGETATION SAMPLING

G. A. McIntyre

I. INTRODUCTION

A grassland research worker is commonly faced with problems such as how much herbage and the proportions of its constituents are present in a plot, field or district, or the percentage ground covered by the various constituents and so on. A complete census in terms of harvesting the whole area or of making observations on all the individuals is commonly impracticable or impossible, particularly where the operation of obtaining observations is destructive, as in the case of yield. It is therefore necessary to resort to sampling; that is, observations are made on some selected individuals only.

We must first introduce some of the terms used in statistics. All the individuals of an area under study or the total area subdivided into distinct units of, say, 1 m^2 each, are referred to as the *population*. Individuals or units selected for observation are termed *sampling units* and the aggregate of such units for the area is called the *sample*. The observations made on the sample provide an *estimate* of the unknown true value of the population. The arithmetical average of the sample unit values is the *mean* ($\bar{x}$). The average value of an estimate from an indefinitely large number of independent sampling units is called the *expect-*

ation. If the expectation differs from the population value, the estimate is said to be *biased* and the difference between the expectation and the population value is the *bias.*

To determine how precise an estimate is, we may calculate the *range* or the difference between the smallest and largest sampling units. A better measure is the *variance,* or its square root, the *standard deviation.* The usual symbol for the population variance is σ^2 which is the expectation of the square of the deviation from the population mean. An unbiased estimate of it can be obtained from the equation $S^2 = \Sigma (x - \bar{x})^2/(n - 1)$. The standard deviation of a mean is called the *standard error* and to determine the degree of precision relative to the mean, the *coefficient of variation,* or 100 x standard deviation/mean may be used, this measure falling as sampling unit size increases, though not proportionately.

Random sampling is the identification of sampling units where all possible units have an equal probability of being selected. Random sampling numbers are often used to identify the units. Some consideration must be given to whether the sample comes from a *finite* or an *infinite* population. In the former and more usual case, no correction need be made if the fraction sampled is small, but if large, the variance of the estimate decreases to the point where, if the whole population is measured, the population value is now known and its estimate has no variance; in other words, it is infinitely precise.

II. DISTRIBUTIONS, PARAMETERS AND STATISTICS

A population of measurements, one from each sampling unit, may be classified according to the magnitude of the measurement into a *frequency distribution.* The measurements may be integral, as with plant counts, or continuous. If continuous, it is best to group the values into 12 to 20 equal intervals to cover the range of measurements. A frequency distribution may have a single peak (unimodal) or several peaks (multimodal). Examples of bimodal distributions occur with perennials which tend to be made up of single-aged individuals because of massive re-establishment following some such disaster as fire, flood or frost. If such a single-aged stand is sharply reduced, it will be followed by an interim mixture of old survivors and young replacements. Most frequency distributions of pasture measurements are unimodal and very often are not symmetrical but have a long tail of low frequencies at high measurements. Such a distribution is said to be *skew positive.* It is to be noted that the shape of the frequency distribution is dependent on the size of the sampling unit. With a very small sampling unit, say of 1 cm^2, in all except very matted vegetation there would be a substantial proportion of units with no material. This would be rare or non-existent with units of 1 m^2, except in arid environments. The general tendency with increasing size is to move from a skew positive distribution to a distribution approaching *symmetry.*

A population of measurements may be characterised by measures of central tendency, spread, skewness, degree of peakedness, etc. These are called *parameters* of the distribution. Estimates of these parameters from random samples are called *statistics*.

The measure of central tendency almost invariably used in pasture measurements is the mean or simple arithmetic average. The expectation of the sample mean is the population mean.

For measures of dispersion, the range is sometimes used as a descriptive measure but the standard deviation σ has substantial theoretical advantages and is commonly employed. The variance is $\sigma^2 = \Sigma\,(x\text{-}m)^2/N$. An estimate of σ^2 from a sample of size n is $S^2 = \Sigma\,(x\text{-}\bar{x})^2/(n\text{-}1)$ where $\bar{x}$ is the sample mean. It is slightly biased, as its expectation is $N\sigma^2/(N\text{-}1)$, evident if $n = N$. For symmetrical natural distributions the limits, mean $\pm\ \sigma$, roughly embrace the central two thirds of the observations.

For measures of skewness, peakedness, etc. average values of higher powers of deviates from the mean are used, but need not further concern us.

It is not appropriate here to go into detail about statistical theory, but there is one relation which should be mentioned. If a random sample unit is drawn from each of two populations and the sum and the difference of the two units is found and this procedure is repeated indefinitely, the variance of the distributions of pair sums and pair differences so formed are both equal to the sum of the variances of the parent populations. From this one can deduce that, ignoring finite sampling, the variance of the sum of n random sample units from a population is $n\sigma^2$, and the variance of the mean of n random sample units is σ^2/n, i.e. the standard error of the means decreases proportionally to the square root of the number of units. Moreover, with increasing sample size, the distribution of the means becomes more symmetrical and bell shaped and in the limit takes the form of a specific distribution known as the *normal distribution*, irrespective of the distribution of the parent population of pasture measurements.

A particular form of distribution is generated if the random sample units are classified as 1 if, say, a particular species is present, 0 if it is absent. Then, if the proportion of units in the population with presence is P, the mean of the population is P and the variance is $P(1 - P)$. With n random sampling units the expected number of presences is nP and the variance is $nP(1-P)$. The frequency distribution of number of presences in samples of size n is known as the *binomial distribution*. The probability of no presences at all in n random sample units is $(1 - P)^n$ and the probability of all presences is P^n.

If a sample unit is divided into k equal parts and the mean number of plants per sample unit is m, then the mean number per part is $p = m/k$ and as k increases indefinitely p becomes very small, with a negligible chance of there being two or more plants in any part, i.e. p becomes the probability of one plant per part. If the probability of presence in any one part is independent of any other part, the number of plants per unit is $m = pk$ and the variance of m is $pk(1-p) = m(1-p) = m$ in the limit. The distribution of the number of plants per unit in these circumstances is known as the *Poisson distribution.* The probability of no plants within a unit is $(1-p)^k = (1-m/k)^k = e^{-m}$ in the limit.

III. SAMPLING OF VEGETATION

The two main fields of investigation involving sampling of herbage are (a) surveys to establish the position with respect to the areas of different types of cover, their productivity and perhaps trends in time, and (b) experimental studies to establish the merits of different species mixtures, managements, fertilizers, etc. Except for short-term feeding trials where the standing feed is an input not affected by treatments, it is almost essential for a sound basis of inference to have replicate plots for each treatment. These may be regarded as samples from the area in which the experiment is set, and variation from sampling within the plots is additive to the replicate variation. The principal objective here is to have enough precision in measurement to establish as significant any treatment effects considered important. This requires that this order of difference be defined and prior information on expected variation between and within plots be available for intelligent planning. There are cases where such information is not available, in which case the experimental effort may best be split into a pilot study to examine the variation, followed by the major experiment. If a trial is to be evaluated primarily in terms of animal products, minimal requirements in plot size and replication may be dictated by variability between animals. This brings up the question of trial objectives, design, set stocking and/or rotation of stock, etc. These issues are outside the scope of this chapter. The main emphasis here will be on sampling within planned experiments. Additional features involved in sampling surveys will follow.

1. Sampling within planned experiments

There are four main approaches to direct sampling within plots, simple random sampling, stratified random sampling, cluster sampling and systematic sampling. In addition, there are methods which use concomitant measures and unbiased selective procedures.

(a) *Simple random sampling*

Simple random sampling, as its name implies, means taking samples at random within a plot, i.e. all possible placements of the sampling unit are given an equal chance of inclusion. To avoid unconscious bias it is best to locate the sites of the sampling units by random co-ordinates on the plot boundaries, the plot being conventionally rectangular in shape. The actual position is normally located by pacing, minor errors in distance being of no importance, except that the length of the pace should not be altered as the target is approached in response to its appearance, and the quadrat should be precisely sited relative to the final step, say at the toecap. With a small quadrat in a variable pasture there is serious opportunity for unconscious bias if any latitude in placement is allowed. In trials which involve sampling of a partially or wholly destructive character through a series of occasions, it may be necessary with small plots to grid the area to provide an adequate number of samples, each with its protective surround for the duration of the trial, or at least until cut areas have had enough time to recover fully. In this case the samples to be taken at any harvest would be located accurately.

The variance of the sample mean is $S^2(1 - f)/n$ where S^2 is the unbiased sample variance ($\Sigma(x - \bar{x})^2/n - 1$) and f is the fraction of the plot which is sampled. If a large area is being sparsely sampled, the factor $(1 - f)$ can safely be ignored.

In addition to the risk of bias from location there can be bias at the quadrat boundary in cutting or counting. With discrete plants, one can count the plants internal to the quadrat and those cut by two adjacent edges and at the included corner but at no other corner. The rule to adopt is that if the quadrat is considered to be part of a grid of quadrats, a plant, or portion of a plant in cutting, will be taken into account only once. Height of cutting can create difficulties. Where changes in time are of most interest the policy of harvesting all above-ground parts is the least controversial. However, particularly with grazing by cattle the interest may be in the amount of plant material potentially available to the stock and this may be arbitrarily set at material above, say, 3 cm in height. If such a rule be adopted it is necessary in planned experiments not to confound cutting teams with treatments within a replicate unless they are using machines with mechanically regulated height of cutting. In any event, it is better to assign a replicate of all treatments to each cutting team or, better still, have each team cut the same number of quadrats on all plots.

(b) Stratified random sampling

Stratified random sampling, in the experimental context, means division of a plot into a number of sub-plots in such a way as to confound environmental gradients with sub-plot differences. This means that the variation within sub-plots will be minimal. The pattern of stratification will be the same in all plots, and except in the unlikely case of the gradient being identical in each plot, this will reduce the gain of the device. In general, some gain may be obtained where points closer together are more likely to be similar than those further apart.

The variance of the mean is $\bar{S}^2(1 - f)/n$ where $\bar{S}^2$ is the average within sub-plots variance (analogous to the error mean square in the one-way analysis of variance), and n is a multiple of the number of sub-plots per plot. If an estimate of the sampling error is required, a minimum of two random units within a sub-plot is necessary. The efficiency relative to simple random sampling is given by $S^2_{plot}/\bar{S}^2_{sub\text{-}plot}$.

(c) Cluster sampling

Cluster sampling is a procedure commonly employed in surveys where the cost of reaching the location of a primary sampling unit is a major expense; it is rarely used within plots. Within such a location, a number of secondary sampling units are taken. One could envisage such a procedure being used in very large plots as in grazing trials with, say, 1 beast to 5 to 10 ha with sub-plots of equal size as primary units. Having located a sample of these primary units at random within the whole plot, secondary samples would normally be taken on a regular pattern within each primary unit. If finite sampling is ignored, the variance of the mean of nm secondary units is S^2_{pr}/nm, where S^2_{pr} is the mean square between the n primary units and m is the number of secondary units per primary unit. With f the sampling fraction for primary units and f′ the sampling fraction of secondary within primary units, the variance of the general mean per plot is $((1 - f)S^2_{pr} + f(1 - f')S^2_{sec})/nm$. This is strictly true only if the secondary samples are randomly located within the primary units. The formula will tend to give an over-estimate if the secondary units are located in a systematic pattern about the primary unit centre. With very large plots or paddocks, some of the devices of survey sampling, such as identifying boundaries of natural strata and using a variable sampling fraction within strata, would be considered.

(d) Systematic sampling

By systematic sampling is meant sampling at regular intervals. In a rectangular plot the simplest arrangement is to take as sample unit sites the central points of the cells of a grid imposed on the plot. For a plane gradient within the plot this will give a very precise mean. This condition would be exceptional but it is common for closer points to be more alike than more distant ones or for the positive correlation between contiguous sites to decrease with distance. If the correlation falls off with distance in a concave upward curve, systematic sampling is slightly better than stratified sampling with one item per stratum and definitely better than random sampling (Cochran, 1963). The primary risk with central point sample units is that there may be some regular natural or imposed features parallel to the plot boundaries, such as past ploughing or seeding irregularity. The risk can be reduced while retaining the advantage of ease of location by sampling along systematically located transects along the width of the area, choosing random sampling points within cells in the first column as starting points. There is some evidence from uniformity data that taking samples at regular intervals along a zig-zag path crossing the field four or five times gives nearly maximum precision with economy of effort for the number of samples taken. The fact that no strictly valid estimate of sampling error of the plot mean is available with systematic sampling is not a handicap in a replicated field trial, unless the information is required for future modification of the sampling intensity and even for this a rough estimate using average variance between adjacent pairs will suffice.

(e) Concomitant measures (Double sampling)

Concomitant measures are often used in pasture sampling to economise in the effort of cutting and drying samples and hopefully to gain in precision by substantially increasing the number of sampling units even though they are measured only indirectly. The general principle is to make the indirect observations (weight estimates, capacitance measurements, etc.) on a substantial number of sites within the plot and make direct measurements on some of them (see chapter 4). The regression of the direct on the indirect measurements is determined and if a straight-line relation is appropriate, the mean of direct measurements for the plot is read off at the mean $\bar{x}'$ of the indirect measurements, $\hat{\bar{y}} = \bar{y} + b(\bar{x}' - \bar{x})$. The variance of the estimated direct mean involves the sampling variance of the indirect mean and the errors of the calibration line.

$$VAR(\hat{\bar{y}}) = S^2_{y.x}((n'-n)/n'n + (\bar{x}'-\bar{x})^2/\sum_1^n(x-\bar{x})^2) + S^2_x b^2(N-n')/Nn'$$

with n' total indirect measures, n the subset of direct measures and N the possible number of samples in the plot. $S^2_{y.x}$ is the variance of residuals about the fitted line and S^2_x is the variance of all the x values, i.e. of all the indirect measurements.

When the indirect measure is objective and highly correlated with the direct measure, the conditions are very favourable to the procedure. However, in order to reduce the effort in direct measurements, there is a temptation to pool observations on direct measures over replicates, within a treatment and perhaps even over treatments as well. If the treatments are such that the aerial distribution of dry matter of individual species varies between treatments, the relation between direct and indirect measures is also very likely to vary and the use of a single relation based on pooled data will then be biased for any particular treatment. The statistical position is further complicated if the regression relation is curvilinear. If reduction of effort is the primary reason for using this procedure, the benefit to be derived can be specified (Wilm, Costello & Klipple, 1944) if the costs of direct and indirect measurements are known. If the costs per observation for direct and indirect measurements are C_s and C_m then to a first approximation double sampling with regression is superior to single sampling if $r^2 > 4C_sC_m/(C_s+C_m)^2$ where r is the corelation betweed direct and indirect measures. For example if C_m is $C_s/10$ then there will be a gain if $r^2 > \cdot 33$ or $r > \cdot 58$ (Cochran, 1963).

(d) Unbiased selective methods

In general, any attempt to select representative sample units is very liable to produce a biased result with no assurance that all treatments will be equally affected. There is a procedure of selected ranked samples (McIntyre, 1952; Dell & Clutter, 1972) which is unbiased and which in favourable circumstances can give worthwhile gain for little extra effort. Suppose three primary points are located at random within a plot and at each of them three quadrats are placed relative to the primary point in some predetermined way. For the first group, identify visually or by some indirect and easy method the quadrat which has the largest measure of interest, for the second group the median measure and for the third the lowest measure. Cut or otherwise determine directly the values for the identified plots. Then, at best, the mean of these cut sampling units will be worth slightly less than six sampling units taken at random. Factors which reduce the profitability of the procedure are: (i) strong gradients in the measure within the plot relative to local variation, since the sets of three contiguous units only sample a local average, and (ii) errors of visual estimate in correctly ranking the quadrats within each. Generally, with k ranks the maximum efficiency is slightly less than $(k + 1)/2$ relative to random sampling. In the presence of gradients within the plot, the procedure would be used most effectively to improve the means of strata within plots. If an error of estimate is required, it is necessary to have replicate sets within plots or strata within plots. The variance of the general mean, ignoring finite sampling, is the average variance of replicates of the same rank within plots or of strata within plots divided by n.

(g) *Number, size and shape of sample units*

In sampling for yield for individual species or the whole pasture, plant material falling within the boundaries of a quadrat is cut and weighed after drying or weighed in the fresh state with a sub-sample taken for drying. Local variation in the field per unit area can be high and on this may be superimposed gradients in production corresponding to changes in soil type, moisture availability, etc. With a rectangular quadrat, the variation between quadrats will be least when the quadrats are directed with the main axis down the gradient of production. Even if a consistent orientation is used, it is likely that, on average, sample units using a long narrow quadrat will be less variable than sample units from square frames of equal area. Furthermore, the more numerous the quadrats for the same total area then, fairly generally, the more precise will be the sample mean, if one can ignore greater risks of bias arising from latitude in quadrat placement and from inclusion or exclusion of plant material on the boundary. The progression to higher precision with diminishing size and compensating increase in number of quadrats, is normally quite regular with dense sown pastures but can show irregularities with cover that is heterogeneous or sparse cover, such as desert shrub. With quadrats much smaller than the size of individual plants, the variance of the mean rises with increasing size up to about the average size of plants; thereafter it falls and will normally again rise due to large scale heterogeneity in production. This implies that it would be preferable to keep the quadrat size in sampling tussock or desert type flora at least several times larger than the average size of the dominant species rather than to reduce it to the average plant size. Another aspect of irregularity with quadrats comparable in size to plants, and which is not uncommon with swards having drilled rows of a perennial grass, is that small quadrats are most effective if the length of one pair of their sides is a multiple of the row width and these sides are consistently oriented across the direction of the rows in placement.

With sown pastures it is possible to give some broad indication of variability in relation to quadrat or plot size. Smith (1938) found from uniformity data that the variance of plot means is related to plot size by the relation $V_x = V_1/x^b$ where V_x is the variance of yield per unit area on a plot of size x. With no correlation between contiguous units, $b = 1$, while with positive correlation b is <1. It can be expected that errors in technique, disease, genetic variability, etc., will tend to increase b. Though the model has certain deficiencies, the relation does describe the data of uniformity trials remarkably well.

There is the further limitation that this form of relation will not be applicable to both a part of a very large area and also to the whole area. This difficulty is overcome if we postulate with Smith that the law applies to the whole area when we may write it $(V_x)_\alpha=(V_1)/_\alpha x^b$. Then it can be shown that the variance of plots of size x in blocks of m plots, $(V_x)_m = (V_1)_\alpha/x^b.m(1 - m^{-b})/(m - 1)$. For sown crops and pastures an average value for b is 0·4. Table 1, based on $b = 0 \cdot 4$, gives coefficients of variation for mature crops and pastures on agricultural land fully utilising the plot area which is square.

TABLE 1.

Table of Coefficients of Variation

Plot or quadrat size in square metres

		0·16	0·5	1	2	4	10	20	40	80	160	400
Quadrats	2	24·3	19·4	16·9	14·7	12·8	10·6	9·3	8·1	7·0	6·1	5·1
per plot	3	25·5	20·4	17·7	15·4	13·4	11·1	9·8	8·5	7·4	6·4	5·3
or	4	26·3	21·0	18·3	15·9	13·8	11·5	10·1	8·8	7·6	6·6	5·5
Plots	6	27·4	21·8	19·0	16·5	14·4	11.9	10·4	9·1	7·9	7·3	5·7
per block	8	28·1	22·4	19·5	16·9	14·7	12·3	10·7	9·3	8·1	7·1	5·8
	16	29·6	23·5	20·5	17·8	15·5	12·9	11·3	9·9	8·5	7·4	6·2
	30	30·6	24·4	21·2	18·5	16·1	13·3	11·7	10·2	8·8	7·7	6·4
	60	31·6	25·2	21·9	19·1	16·6	13·8	12·0	10·5	9·2	7·9	6·6
	120	32·4	25·8	22·5	19·6	17·0	14·1	12·3	10·8	9·4	8·1	6·8
	240	33·0	26·3	22·9	19·9	17·4	14·4	12·6	11·0	9·5	8·3	6·9

Generally speaking, the values in this table are 10 to 15% higher for cereal grains even allowing for shape of plot. The effect of shape will depend on the direction of the fertility gradient, but as a rough guide the effect of changing the length/breadth ratios from 4×4 to 1×16 would reduce the coefficient of variation by a third. Disease and pests will, in general, increase the variability. The relative variability also decreases as the yield per unit area increases due either to fertility or to maturity. With grazing, the animal can impress the selectivity of its grazing on the natural variation with variable effect, depending on the nature of the pasture and the intensity of grazing. For continuously grazed phalaris-subclover pasture at Canberra, the coefficients of variation for quadrats of about 1 m^2 in large plots were as follows :

Amount present (dry) kg/ha	Coefficient of variation
1	42
10	40
20	34
30	28
40	23
50	20
60	17

The values of Table 1 may be used to give quantitative illustration of points made earlier. Suppose we have a randomised block design with 6 treatments and plots 160 m^2 in size. Then with mean of 100, the between-plot variance within blocks is $7 \cdot 3^2$ or 53, if the plot is fully harvested.

With 2 quadrats of 4 m^2 at random, the sampling variance of the plot mean is $16 \cdot 3^2 \times (1-8/160)/2$ or 126 so the between-plot variance is $53 + 126$ or 179. Here $16 \cdot 3$ is an interpolate in the table for quadrat size of 4 m^2 with 40 quadrats per plot.

With 8 quadrats of 1 m^2 at random, the between-plot variance is $53 + 22 \cdot 6^2 \times (1-8/160)/8 = 53 + 61 = 114$.

With 8 quadrats of 1 m^2, two at random in each quarter of the plot, the between-plot variance is $53 + 21 \cdot 5^2 \times (1-2/40)/8 = 53 + 55 = 108$.

With 8 quadrats of 1 m^2, 2 from each quarter of the plot, selected by ranking, if fully efficient, the between-plot variance is $53 + 55 \times 2/3 = 90$.

With plots of 8 m^2, fully harvested in the same design with 6 treatments, the between-plot variance is $13 \cdot 0^2 = 169$.

These examples illustrate the advantage of small samples for the same total area sampled; of stratified against simple random sampling; and of selective sampling. The last example shows that greater precision is possible using samples within large plots than fully harvesting small plots of the same area as the total sample.

(*h*) *Subsampling for dry weight*

Savage (1949), reporting on grass species, strains of sweet clover and of brome grass, weighed green bulk in the field together with subsamples of 0·5, 1 and 1·5 lb green weight. These were subsequently dried. The coefficients of variation of dry matter per cent were:

	0·5 lb	1·0 lb	1·5 lb
Grass species	2·89	2·50	2·04
Sweet clover	5·06	4·87	3·46
Bromes	4·62	2·87	—

Grassland Research Institute, Hurley (1961), reporting on subsampling for dry matter, with about 1 kg weight of green subsamples, found a coefficient of variation of about 4 in cuts from a winter pasture which was often very heterogeneous.

At Canberra drying percentages for lucerne strains using a 2 lb sample gave a c.v. of 6·4 on the average.

If dry weight is 100, the effect of variation in drying percentage is to make an addition to other sources of variation of $(\text{c.v.})^2 \times (1-f)$ where f is the subsample fraction.

(i) Estimation of component species

The data in Table 1 relate to the total yield. With high aggregation of species, the advantage of many small quadrats over a few large quadrats for yield of particular components would be accentuated. Commonly, an analysis by species is done through subsampling of the bulk cut from a quadrat (see Chapter 3) and this may give rise to further serious errors if instead of the whole subsample, only a fraction of it is separated. The magnitudes of these errors are specific to the species mixture and technique of separation, as green or dry, partial separation of long stemmed species before separation of the finer material and so on. In particular circumstances it may be an economy to use a calibrated rating, ranking or weight estimate system of standing pasture, e.g. the dry-weight rank method of 't Mannetje and Haydock (1963), or even to take a larger independent set of quadrats from the field, especially for botanical analysis (see Chapter 3).

(j) Sampling intensity in relation to replicate variation

In a designed experiment, say randomised blocks, the primary interest of the experimenter is in the comparison of treatments and his reference for uncontrolled variation is the variation between replicate plot means. If the plot has been measured by samples, the variation between plot means has two components, the variation between means of fully harvested plots plus the variation between the possible sample means and the true plot mean. The analysis of the observed plot means will furnish a direct measure of this composite variance and the only virtue of having information on the contribution from sample variation within the plot is to assess the consequences of altering the sampling intensity for any future occasion when circumstances may not be very different or, more rarely, where differences in variability within plots induced by treatments may be of direct concern. At most harvests there is little interest in variability within the plot so that the deficiencies in estimating the sampling variance of systematically placed sample units do not operate against its use. It is important, of course, that the variation of the mean from sampling within plots should not be substantial in relation to the variation between plots that are fully harvested. The composite variance $\sigma^2 = \sigma_p^2 + \sigma_s^2$. With $\sigma_s = 0 \cdot 5\sigma_p$, then $\sigma = 1 \cdot 12\sigma_p$. With $\sigma_s = 0 \cdot 25\sigma_p$, $\sigma = 1 \cdot 03\sigma_p$, which is a poor return for an approximately 4-fold increase in sampling intensity. To revert to an example given earlier, with 8 quadrats of 1 m^2, 2 from each quarter selected by ranking, the sampling contribution to the composite variance is 37 and $\sigma = (90/53)^{1/2}\sigma_p = 1 \cdot 30\sigma_p$. The c.v. of plot means is $100 \times /90^{1/2}/100$ or $9 \cdot 5$, in contrast to $7 \cdot 3$ as the limit with no sampling error. The percentage standard error of a treatment with r replicates is $7 \cdot 3/r^{1/2}$ at best, or $9 \cdot 5/r^{1/2}$ with this level of sampling variation within plots. This may be adequate precision for the comparisons of interest; if not, there is little to be gained by further reducing the contribution for sampling variation within plots.

To summarise, the primary aim of sampling within experiments is to provide unbiased estimates of interest to the experimenter with precision adequate to establish differences regarded as important and with minimum effort and cost. To achieve this objective, it is necessary to define what minimal differences between treatments it is desirable to establish and to have some foreknowledge of the natural variability between replicates and between sample units within replicates likely to be encountered. Given these, it is not difficult to devise a design and sampling program which has a very reasonable chance of attaining these objectives. The price of achieving the required level of precision may be prohibitive, but it is better to know this beforehand than to proceed with a programme which cannot be realized satisfactorily.

2. Survey sampling

In general, the procedures in survey sampling are more complex than in sampling in planned experiments, because the natural variability is greater and the area to be covered and the associated difficulties of reaching sampling sites are very much greater. This puts more emphasis on sampling devices that will give economies of effort without introducing an unacceptable level of bias. There are many excellent texts covering both the statistical and practical aspects, e.g. Cochran (1963), Yates (1953), Sampford (1962) to name only a few, as well as very numerous journal articles on sampling in specialised fields. Sampling for ecological purposes is discussed by Greig-Smith (1964).

In what follows the main features in survey sampling which are additional to those met in sampling in planned experiments will be touched on only briefly.

(a) Stratified sampling

It may be possible by ground survey, or by the use of aerial photography and photo interpretation based on adequate ground transects, to map the area of study into strata roughly homogeneous with respect to vegetative cover. These strata may then be sampled with randomly located plots from which means and standard errors for strata may be determined. Mean estimates for the whole study area may then be obtained by combining data from the strata, provided these are weighted according to the proportion of each stratum of the whole area. The intensity of sampling is not necessarily uniform but is usually a compromise between having a satisfactory minimum level of precision for each stratum mean, greater representation of the more variable strata and of strata which can be most economically sampled if cost is a consideration.

(b) Cluster sampling

This is a sampling method which generally is not efficient per sample unit observed but may give an estimate of the population mean with a prescribed level of precision at least cost. The population is divided into primary, or first stage, units and secondary, or second stage, units. The partitioning may proceed to further stages. The selection of the samples at each stage and the estimate of the population mean, given that the size of all sampling units at each stage is known, allows many possibilities in sampling procedure, each with its corresponding estimate of error. One of the simplest methods is selection with replacement of primary units in proportion to size. The reader is referred to texts on survey sampling for more detail.

Cluster sampling may be used in conjunction with stratified sampling to reduce the cost within strata. Analytically, the strata are populations as far as estimates from the cluster sampling are concerned. The general population mean is then obtained from the weighted strata means.

(c) Systematic sampling

This is most likely to occur in forest or dense shrub where evenly spaced transects give a simple procedure for identification of vegetation types within boundaries visible in an aerial photo and sample plots at regular intervals give samples within these identified strata. They are not random plots, but the degree of bias in the estimate of error is likely to be quite small. The estimate of the variance of the general mean from the weighted strata means will have very little bias and the whole procedure has a good prospect of being efficient in terms of labour expended.

To summarise, the primary objective in survey sampling is to obtain estimates of means, proportions, variability, etc. within some defined level of precision and with least cost or effort. It is a matter of selecting those procedures most likely to be effective in respect to bias and reliability in terms of the pre-knowledge of the population, the cost of transport and measurement and other limitations of resources.

IV. REFERENCES

COCHRAN, W. G. (1963). Sampling techniques. New York: John Wiley & Sons, Inc. 2nd Edn.

DELL, T. R. & CLUTTER, J. L. (1972). Ranked set sampling theory with order statistics background. *Biometrics*. **28,** 545-55.

GRASSLAND RESEARCH INSTITUTE, HURLEY (1961). Research techniques in use at the Grassland Research Institute, Hurley. *Commonw. Bur. Pastures Field Crops, Hurley, Bull.* 45.

GREIG-SMITH, P. (1964). Quantitative plant ecology. London: Butterworths. 2nd Edn.

MCINTYRE, G. A. (1952). A method for unbiased selective sampling, using ranked sets. *Aust. J. Agric. Res.* **3,** 385-90.

MANNETJE, L. 't & HAYDOCK, K. P. (1963). The dry-weight-rank method for the botanical analysis of pasture. *J. Br. Grassl. Soc.* **18,** 268-75.

SAMPFORD, M. R. (1962). Introduction to sampling theory with applications to agriculture. Edinburgh: Oliver & Boyd.

SAVAGE, R. G. (1949). Moisture determination in the comparative testing of forage crops for hay yield. *Sci. Agric.* **29,** 305-29.

SMITH, H. F. (1938). An empirical law describing heterogeneity in the yields of agricultural crops. *J. Agric. Sci.* **28,** 1-23.

WILM, H. G., COSTELLO, D. F. & KLIPPLE, G. E. (1944). Estimating forage yields by the double-sampling method. *J. Am. Soc. Agron.* **36,** 194-203.

YATES, F. (1953). Sampling methods for censuses and surveys. London: Griffin. 2nd Edn.

Chapter 3

MEASURING BOTANICAL COMPOSITION OF GRASSLANDS

J. C. Tothill

I. INTRODUCTION

Differences in environments as caused by climate, topography, geology or soils and interference such as grazing, watering, fertilizing and introducing plant species result in different vegetations. The aim of measuring botanical composition of grasslands is to describe species composition or to monitor changes in composition. The level at which botanical composition is determined depends on the objectives of the research.

The term ' botanical composition ' implies gathering botanical information about areas of vegetation so that they can be organized into some coherent framework which is pertinent to the purposes of the study. Although botanical composition has tended to be presented mainly in descriptive terms, detailed quantitative studies of herbaceous vegetation have been conducted for many years. For example, in New Zealand Cockayne (1926) initiated the use of a point quadrat for measuring floristic cover on native grassland; later Levy and Madden (1933) formalized the technique by using pins mounted in a frame in order to follow changes in the composition of simple grass/legume pastures, while De Vries (1933) studied specific frequency on mixed pastures in the

Netherlands.

Recent developments in the field of data handling, following the introduction of computers, have given tremendous impetus to the development and use of quantitative methods, and have also produced some powerful means of handling formalized qualitative and semi-quantitative data (Williams, 1971, 1976; see also Chapter 8). Not only is it now possible to examine larger tracts of vegetation in greater detail, but more information about the vegetation under study can be incorporated into the analysis.

This chapter aims to build on Brown's comprehensive review published in 1954, though of course it will be necessary to overlap it in some cases, and will also take account of more recent reviews and bibliographies in the field since 1954, e.g. Cain and Castro (1959), Phillips (1959), Anon (1959, 1963), Tothill and Peterson (1962), American Society of Range Management (1962), Greig-Smith (1964), Norris (1967), Shimwell (1971), Anon (1972).

II. DESCRIBING AND MEASURING VEGETATION

The processes of describing and measuring vegetation are usually carried out simultaneously, but in fact are two different operations. It should be borne in mind that description is not used here in the sense of an entirely qualitative process, nor is measurement used in an entirely quantitative sense. In describing vegetation we are essentially seeking either for pattern, which is a static description of what is there, as required in surveys, inventories etc., or vegetation changes, which are time-dependent studies of sequential changes in patterns or of quantities of components, as in studies of factor effects on vegetation. The processes of recognising and describing patterns may be largely intuitive (i.e. the recognition process is not clearly defined) or it may be based on systematic quantitative or qualitative measures. In such cases, the description need only portray the vegetation at a degree of precision no more than that appropriate to the particular study. On the other hand, where it is wished to follow vegetation changes over time, a relatively high level of precision of vegetation description will be needed.

1. Scale in relation to sampling in vegetation analysis

The initiation of any study of vegetation requires a prior decision as to the most appropriate scale. Since it is usually not possible to study vegetation in its entirety, some form of representative sampling is usual. The sampling procedure to be followed depends on both the scale and nature of the approach to be taken, i.e. whether it is a large-scale inventory for the purposes of mapping, pattern analysis on medium to small scales, static or time sequence studies, or studies directed at specific aspects of vegetation behaviour and performance. It is necessary to consider what the main constraints on sampling might be, particularly in relation to the resouces available and to the area that must be studied.

Other considerations are the level of accuracy required, the size and number of the samples to be taken, their distribution, type and so on.

(*a*) *Large scale*

This is usually the scale of broad surveys on a regional or even continental basis. They may be made for inventory purposes or comparative vegetational studies. Because the areas involved are invariably large, rapid methods are usually adopted, keeping visual estimation at a maximum and actual measurement of quantitative characteristics at a minimum. The methods used may be applied over the area as a whole or by extrapolating from carefully documented reference points. The qualitative nature of the measurements in these cases, however, can be quite reliable if they are formalized by means of coding; of course, the experience of the investigator also counts a great deal.

In recent years, this type of analysis has been helped considerably by remote sensing at various scales. These methods have not yet found a great deal of direct application to vegetation measurement, though they have to survey and mapping and in the identification of areas for closer study (Driscoll & Francis, 1970; Driscoll *et al.*, 1970). Greater use of large-scale aerial photographs is being made (Carneggie, Wilcox & Hacker, 1971; McCown, Tolson & Clay, 1973) and experience in photo-interpretation has allowed investigations over much larger areas and in greater detail than was previously possible. The recognition of a recurrent pattern on the aerial photograph which can be typed and verified from selected ground study sites permits accurate extrapolation over large areas as well as the identification of areas for closer study. However, maps on a continental scale, such as that of Rattray (1960) are based on all sources of information, e.g. existing maps, aerial photographs, rainfall or climate maps and collected experience. The scale of pattern sought is that associated with differences in large-scale topography (mountains, plains, valleys), rainfall, geological or great soil groups and even latitude.

At this scale a physiognomic-structural classification is normally used, though if it is a self-contained genetic region, the predominant floristic components may also be included (see next section for a discussion of this terminology). In the case of Rattray's (1960) map of African grasslands the analysis is largely restricted to only one physiognomic-structural part of the vegetation (i.e. the herbaceous layer) and the main floristic components must be used as descriptors. Moore's (1970) treatment of Australian grassland vegetation follows similar lines.

(*b*) *Medium scale*

This may be considered the scale at which vegetation is normally investigated intensively. Even if quite large areas are involved it is possible to undertake considerably detailed ground studies. In agronomic work it is the experimental area or paddock size area which is being considered. The scale of pattern sought is that related to differences in soils, medium-scale topography (ridge, slope, hollow, aspect), soil moisture (shallow and deep soils, drainage) and past historical and present factors (cultivation, clearing trees, burning, fertilizing, grazing, fence lines).

Invariably, investigations at this scale are carried out on a floristic basis although sometimes it may be on floristic groups (sown grasses, other grasses, legumes, weeds, forbs, sedges, etc.). Whereas in the case of large survey-scale investigations the classification or summary is usually presented in the form of a map, this is not always the case here, particularly in agronomically oriented studies. On the one hand the search for patterns in natural vegetation usually stems from efforts to elucidate environment-vegetation cause-and-effect relations, in which case maps can be useful, particularly if other factors such as topography and soils, can also be portrayed or superimposed. On the other hand, agronomic studies of vegetation are concerned with changes which may take place in the botanical composition or vegetation pattern as a result of certain treatments or factors imposed. In these cases, presentation of the data in tables or figures is preferable.

(*c*) *Small scale*

In recent years, microscopic examination of plant fragments (Plate 4) from the digestive tract or in the excreta of animals has been used increasingly to examine the floristic composition of the eaten components of different vegetations (see Chapter 7). Not only is it of interest to agronomists to know what plants domestic livestock are eating in different vegetation systems, at different times of the year, or what degree of dietary overlap there is between different classes or species of animal, it is also of great importance to understand the habitats of wildlife adequately in order to prevent their deterioration or to aid their rehabilitation. Plant fragments in animal faeces (e.g. Croker, 1959; Storr, 1961, 1968; Martin, 1964; Stewart, 1967; Stewart & Stewart, 1971; Hansen & Ueckert, 1970), oesophegeal samples (Heady & Torell, 1959; Davis, 1964; Leigh & Mulham, 1966a, b; Galt *et al.*, 1968, 1969; Williams, 1969), rumen or stomach samples (Chippendale 1962, 1968) may often be used effectively to identify dietary components. Most of these applications have been largely qualitative, but Sparks and Malechek (1968) and Free, Hansen and Sims (1970) have had satisfactory results in predicting the dry weights of the species comprising more than five percent of the diets sampled from either faecal or oesophageal material. This was accomplished by grinding the material to pass a 0·5-mm screen, which reduces problems associated with differences in particle size. Since the fragments are dispersed at random on the microscope plate, relative density can be computed from frequency estimates through the relationship suggested by Curtis and McIntosh (1950) (see also discussion of frequency sampling later). Relative

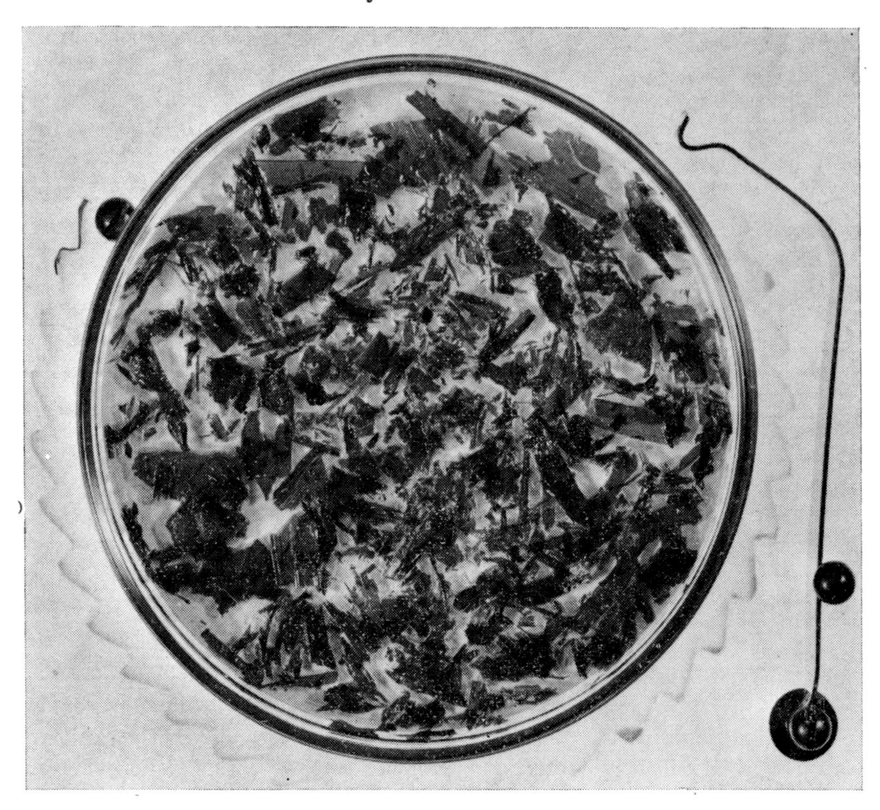

PLATE 4. Plant fragments from oesophageal fistula sample in petrie dish for microscopic examination of botanical composition

density was found to correlate closely with percent weight for each species.

The prospect of differential rate of passage of some feed components through the animal has been considered by various workers; e.g. Free, Hansen and Sims (1970) had difficulty with some of the smaller rangeland forbs. Slater and Jones (1971) found that white clover was apparently completely eliminated by the digestive process; although this differed from the results obtained by Hercus (1960) with white clover and may therefore be a matter of technique. Subsequently, it was established that white clover cuticle resists digestion but in preparation becomes refractively indistinguishable from the mounting medium. However, it can be observed if it is treated with the lipid stain Nile blue (Birk, 1972).

2. The properties of vegetation

The properties of the vegetation are the qualities, characteristics or attributes which may be ascertained and which describe the patterns and processes to be studied. There are four such properties used in vegetation description and measurement, namely *physiognomy*, *structure*, *function* and *composition* (Fosberg, 1967; Shimwell, 1971). They are by no means mutually exclusive though they do represent very different aspects of the vegetation and therefore have different applications, either singly or together.

These properties fall naturally into two fundamental divisions; the first comprises physiognomy, structure and function, the second, composition. The first division is clearly independent of the second and takes account of the fact that similar types of environments, through the process of parallel evolution, tend to evolve similar plant forms and vegetation types. This has been recognized by geographers for a long time and is an important parameter in Köppen's world climate classification. In comparative vegetation studies on an intercontinental or world scale, these are the properties with the most relevance. The second division encompasses the concept of floristic composition as a basis for description. The use of this concept implies that the whole of the region over which comparisons are to be made has had the same genetic resources available to it. In practice, vegetation studies organized in this way are often related to a more generalized physiognomic-structural scheme to enable comparisons to be drawn with other areas of vegetation.

The choice of the most appropriate property to study depends in part on the scale of the investigation and in part on the relevance of the property to the investigation. Thus a clear idea of what aspects of the vegetation to which the property relates is required.

(*a*) *Physiognomy*

Physiognomy concerns the appearance of the vegetation as determined by the life-form of its dominant component. Thus we have forest, woodland, scrubland, heath, grassland, herbfield, etc. These are all terms in common use, and although it may be necessary to define the boundaries for greater precision, there is no great difficulty in doing this and their concept is uncomplicated, i.e. they are qualitative terms but clearly definable. However, because they are very broad in their definition the unqualified use of physiognomy leads to very general classifications. Because of this, physiognomy is generally linked with structural and functional properties to allow greater definition of variables. Webb (1959) has given a physiognomic classification of Australian forests.

(*b*) *Structure*

This is a strictly morphological concept comprising arrangement, spacing and size, the various aspects of which can be expressed in quantitative terms. Structural features of the components of vegetation are height, stem size, crown size, coverage and density, branching habit, canopy layering or stratification and the attendant heights and spacings of the plants within them. Structure alone is not usually employed in describing vegetation, but is combined with physiognomy to produce physiognomic structural classifications or descriptions. Recent samples are those of Specht (1970) for Australian vegetation, and Webb *et al.* (1970) for Australian rain forests. Grabau and Rushing (1968) have given a quantitative, computer-compatible system for general physiognomic-structural classification.

Structure may also be measured in conjunction with composition, since even within grasslands the structure of the vegetation can change rapidly and dramatically in response to changes in grazing intensity, with the consequent longer-term changes in life-form with changes in composition (Duffey *et al.*, 1974).

(*c*) *Function*

Function may be defined as those features which seem to suggest special adaptation to environmental situations, e.g. periodic phenomena such as deciduousness or evergreenness, life-forms such as annual or perennial, growth-forms exhibiting resistance to fire (pyrric), drought (xeromorphy, sclerophylly), water (hydromorphy, mesophytic), temperature (temperate, tropical, boreal) and so on. Function is also used in conjunction with physiognomy and structure and can only be expressed in qualitative terms.

(*d*) *Composition*

The floristic composition of vegetation immediately introduces a more detailed level of investigation. Describing vegetation in floristic terms does not allow comparative studies of vegetation outside the region in which the same species occur. However, it is only at this level of detail that small-scale patterns can be detected, and this is the scale of most vegetation studies, either strictly ecological or agronomic. Most of the discussion of this chapter will therefore deal with vegetation analysis at this level.

(*e*) *Mixed properties*

There appears to be some confusion in the literature about the precise definition of the properties just discussed, but this is probably because they are confounded in varying degrees by different investigators. The way in which these criteria are used depends much on the scale of the study, either in terms of area or in detail, and this is very much a product of the scope of the work. Küchler (1967) has made a plea for greater uniformity in the use of these terms to help in comparative studies of vegetation. It may generally be said, however, that physiognomic-structural classifications are more suitable for large-scale regional or continental investigations than classifications based on floristic composition, since they have a common conceptual and linguistic basis (Fosberg, 1967). Webb *et al.* (1970) pointed to the rapidity and ease with which physiognomic-structural classifications could be carried out, since they did not require a knowledge of the flora, which is a considerable obstacle in rain forest studies, and much could be done by inexperienced observers. Köppen's world classification of climates (see Trewartha, 1954) is a good example of a physiognomic-structural-functional classification, the qualitative aspects of which have been symbolized and thus become amenable to analysis.

3. Qualitative and quantitative values of vegetation properties

The choice of the appropriate property to study depends on the scale of the investigation and on the relevance of the property to the investigation. But the choice of value must also be considered. If the scale is such that only simple and rapid measurements can be made, the investigator is limited to very few measures. However, if scale is not a limiting factor, the questions asked are : What can be measured and what measure is most relevant to the study ?

Not all the properties of vegetation have precisely measurable values. With physiognomy the terms are broadly descriptive, and with function, still descriptive but functionally qualified. Structure and composition are largely described by quantitative or objective qualitative values, all of which can be measured fairly precisely.

Values may be considered at two levels. Primary values are those that can be measured directly, such as number, weight, cover, size, height, length, depth, spacing, presence and so on and secondary values are those which, by means of correlation or regression analysis, can be used to estimate primary values, e.g. chlorophyll content can be used to estimate proportion of green material, spectral reflectance can estimate cover or structure.

Since the primary concern of this chapter is with fairly detailed studies of vegetation and with herbaceous vegetation in particular, attention will largely be addressed to the property of composition and the various values by which it may be measured.

4. The values intrinsic to floristic composition

Absolute values of composition (Greig-Smith, 1964) are *number* (density), *cover* (area) and *weight*. They are independent of the mode of sampling since they are measured directly and expressed in relation to unit area. *Presence* (observed frequency of occurrence), on the other hand, is non-absolute since it has no directly measurable dimension and is dependent on the mode of sampling, i.e. increasing the sampling unit size increases the value of the measured presence of a species. The importance of this distinction is that number, cover and weight measurements are comparable in time and space, whereas presence measurements can only be compared if the sample size and number are known.

It is important to understand what aspects of composition these values actually express. It is likely that one of these values is distinctly more appropriate to the particular study than another. Thus, if the primary objective of the investigation is to obtain an estimate of the composition of pastures in assessing animal production, weight is the most suitable value to measure. On the other hand, if the study concerns rainfall interception or photosynthesis, then cover may be the more appropriate. Number is likely to be the most appropriate value measured in studies of seed germination and establishment or demography. Each of these three values can provide information on presence or frequency, but usually not the reverse, except the special case of frequency being logarithmically related to number when the population is distributed at random, which is usually not the case (Greig-Smith, 1964). Presence (frequency) measures dispersion, distribution or pattern, and when applied over a time period can measure changes in pattern or occurrence. Used in this way it is insensitive to subtle changes in botanical composition, but this is often not important, particularly in fairly large-scale or long time-sequence studies where the differences being measured are large.

A further consideration is that of the time and effort that can be afforded. If the area being investigated is extensive, or alternatively rather variable and requiring intensive sampling, consideration must be given to the ease and rapidity of measuring the appropriate value. Presence (frequency) sampling has the advantage of objectivity and rapidity and for this reason is often used, even where another value might be more suitable. Estimations or secondary measurements may help to speed up the measurement of some of these values but they also reduce accuracy. However, quick estimates of one or more values may often be incorporated with presence sampling, thus combining the objectivity and speed of the latter with some estimate of another dimension. This may often provide an advantageous broadening of the usefulness of the investigation.

In practice, this has been used by a number of investigators who have formulated various combination indices such as the ‘ importance index ’ of Curtis and McIntosh (1951). In some cases part of the information is measured and part estimated, as in the early work of Raunkiaer (see Brown, 1954; Greig-Smith, 1964), where species presence was combined with ‘ frequency symbols ’ or estimated numerical ratings. Another example is the ranked frequency of De Vries (1933) (see also De Vries & De Boer, 1959) where the species are ranked in order of their respective bulk or dry weight. The dry-weight-rank method of ’t Mannetje and Haydock (1963) is a more recent adaptation of this approach.

(*a*) *Number*

Number is often expressed as the density or abundance of a component and represents the count of one or all of the species or parts of plants such as tillers, flowering tillers, vegetative tillers, etc. in a given area. It is a strictly quantitative measure and usually expressed as number per unit area. In this context, and unlike frequency, in comparative studies it is not complicated to the same extent by the size of the sampling unit. Brown (1954) preferred the use of the term ' number ' for this value because of the problem arising from the misleading colloquial use of density and abundance related to cover. Curtis and MacIntosh (1950) used density for number and defined mean area as the reciprocal of density, while abundance was defined as a function of density percent and frequency. Cain and Castro (1959), however, used abundance in the sense of estimated number, usually by forming abundance classes, while they defined density as the counted or measured number of individuals.

Number is obviously a suitable measure where the individuals are clearly distinguishable and it has therefore been widely used in forestry. However, in much grassland work the individuals are often not clearly distinguishable, except in newly established or open or bunch-type grasslands, and the method has limited application. A further problem arises where the population comprises some large individuals, e.g. bunch grasses, and very small individuals, e.g. sparingly tillered annuals, and even stoloniferous or rhizomatous plants. Each of these components is not strictly comparable in terms of number on morphological grounds. As the number of species increases, number becomes more difficult to measure. The usual method of measuring number is by means of quadrat counts.

As a result of the above difficulty of clearly and quickly defining plant units in grasslands, number is often estimated (' abundance ' of Cain and Castro). The estimation of numbers in stands of plants of mixed form is very difficult and has often been found quite erroneous. Raunkiaer's law of Distribution of Frequency Numbers (see Greig-Smith, 1964) demonstrates that there are more rare species than common ones but the dominant species, being the most obvious, is usually overestimated. Such other features which attract the attention of the observer as broad leaves, flowering and ready recognition also tend to cause overestimation. Where large form or habit differences exist in the individuals of a vegetation, it is better either to stratify the sampling into categories or class ranges, if practical, or to look for some other attribute which expresses these differences realistically and without bias, e.g. cover or biomass.

The widely used Braun-Blanquet System of recording vegetation (Kershaw, 1964) incorporates two scales combining abundance and cover of a species as well as an indication of grouping. Nested quadrats are used to compile species lists (frequency) on a minimal-area basis, with abundance, cover and sociability estimated by eye.

Measures of abundance generally bear a close relationship to cover, but because of the effects of variation in growth form of different species, the relationship is not consistent.

Number applies well to studies of early developmental stages and establishment of seedlings such as percentage germination, rate of tillering, persistence of individuals in strain trials or sown pastures where small differences in growth form exist or have not yet developed, when only one or a few classes of individuals in the vegetation is of interest to the study, such as weed control effectiveness, weed infestation, disease or the incidence of other disorders. The distance between individuals in a stand is inversely proportional to the density or number and this relationship has been developed by Cottam and Curtis (1956) and others (Cain & Castro, 1959) as a measure of density. It has also been applied to bunch grasslands by Dix (1961) and others (see discussion of plotless sampling later). The method obviously has limitations with sward forming or stoloniferous /rhizomatous grasses.

(b) Cover

This is the proportion of ground or area covered by the whole vegetation or by individual species as measured by the perpendicular projection of their aerial parts onto the ground (Brown, 1954; Greig-Smith, 1964; Kershaw, 1964) and is usually expressed as a percentage. Since vegetation is seldom disposed in a single plane it is usually possible in closed or near-closed swards to have more than 100 percent cover through overlapping leaves and stems if it is based on individual species. Any sampling procedure must take into account either all of these overlapping parts or only those parts initially encountered, i.e. total canopy cover or cover by the upper stratum. Also, because the size of the aerial part changes considerably in relation to growth during the growing season or as a result of grazing, cover may be expressed by the less variable characteristic of basal cover, i.e. the area occupied by the plant bases. Cover is often a good measure of abundance and is widely used in this sense when, as is frequently the case, it is impossible to give exact definitions to the plant units.

The usual method of measuring cover is by means of point samples, a procedure which is discussed more fully later (see page 44). The impossibility of achieving a point of zero mean area, the problems of locating such a point in the vegetation without bias, the effect of wind on the stability of herbage and the difficulty of avoiding disturbance of the herbage in sampling are all factors affecting the accuracy of the estimate. The method is also tedious if carried out on a reasonably large scale.

Cover may also be estimated by measuring the intercept of the cover component (canopy or basal area) along a measured transect. Although generally faster, the method requires the definition of the plant unit intercepting the line which the point sample avoids. Fisser and van Dyne (1966) examined the influence of number and spacing of points on the accuracy and precision of basal cover estimates by the line-intercept method on Montana foothill range. Sod forming species were sampled best by random placement of points, bunch grasses by systematic placement and species with high basal cover were sampled equally well by either system of placement.

Because of the tedious nature of measuring cover, it is often estimated by eye, but such estimates are subject to the same errors and biases involved in the eye estimation of number. For broad-scale surveys where an estimate of cover is desired, cover classes can be established so that broad differences in cover can be detailed. Morris (1973) has reported the use of rated microplots in estimating aerial and basal cover. With trained operators, the method is five times as rapid as point quadrat analysis and almost as accurate.

Bare ground, or the absence of cover, may be an important feature of the vegetation. The amount of bare ground present in an area of natural vegetation usually reflects the severity of the climate, although it may be due to local micro-climate, soil chemical and physical factors and so on. In managed vegetation it can measure the effects of overgrazing and provide an important guide to soil erosion potential, rainfall interception and infiltration capacity. Cover, on the other hand, is an important measure of the effects of grazing management, fertilizing, burning and growth, particularly when partitioned between the main species components of the pasture.

While it is difficult to return to the same points when point sampling, charting or photographing may be undertaken on fixed positions (Brown, 1954), Goodwin & Walker, 1972). The cover can then be directly measured, though it is difficult to portray the multidimensionality of a tiered situation. It does provide a very accurate record in following changes in actual basal area. Tueller *et al.* (1972) have combined measuring cover from the integrated use of oblique and vertical ground photography and aerial photography with presence sampling for range condition determination.

(*c*) *Weight*

The biomass or dry matter of the individuals or groups of individuals making up a vegetation is an objective measure of the vegetation in terms of relative productivity or percentage composition by weight. It is widely used in agronomic studies of pasture systems, but less so in natural pasture systems of complex floristic composition. There are several methods to determine composition by weight, some very laborious, often requiring hand separation of the different components. Weight data may be related to cover estimates if the total number of hits on each species is recorded in point sampling for cover, provided the growth forms of the species components do not differ greatly. Heady and Van Dyne (1965) have developed a method which predicts the composition by weight using point sampling on clipped herbage.

Weight is perhaps the value most amenable to mechanized measurement and a great number of mowing and cutting techniques are in use (see Chapter 4). Mostly they have the disadvantage of spoiling the herbage for any separation of its components. Consequently, methods have been developed which estimate the proportions of the components in the sample prior to cutting and these are reviewed by Brown (1954). A more recent development from an older approach by De Vries (1933) (see also Dirven, Hoogers & De Vries, 1969) is that of 't Mannetje and Haydock (1963), whereby the first three components of yield are ranked and, through a set of multipliers which have been empirically determined, percent composition by weight is calculated. The actual dry matter yields of the components can then be calculated from the plot dry weights determined by yield sampling (Chapter 4).

Indirect methods of estimating percent composition by weight have been used by Jones and Evans (1959) who derived a satisfactory relationship between cover (as measured by point sampling) times measured height of the herbage and dry weight. Olusuyi and Raguse (1968) derived a relation directly from the first three hits on each thrust of an inclined point. They concluded that the method was not necessarily faster for accurate estimates, but it was non-destructive.

(d) Presence

Presence is usually bracketed with absence. It may alternatively be termed occurrence, existence or frequency. However, frequency is a special statistical expression of presence since it describes the probability of finding a particular species in an area or, as frequency percent, the percent of samples in which a particular species occurs.

Presence is perhaps the simplest value to record, since it requires only the recognition of the occurrence of a species in the sampling unit. Although qualitative, this is a very objective determination, only providing difficulty of measurement when the definition of a plant unit is unclear. This has led to the qualification of frequency being measured as either rooted-frequency or shoot-frequency, the former where only species with their bases within the sampling unit are recorded and the latter where any plant material within the sampling unit constitutes an occurrence. Unqualified, frequency is usually understood to be rooted-frequency and as such contains information on basal area and number or density, though it cannot be precisely equated with these values. However, shoot-frequency may also be used and this contains information on cover.

As presence or frequency gives information on the dispersion or distribution of species, it is the most useful value in ascertaining pattern (see Chapter 8 and Williams, 1976).

There is a very general relationship with abundance, although not a formal one, but generally it may be said that species with high frequencies are also abundant. However, if the evaluation must be on a broader basis and expressed quantitatively, the frequency-density index of Acocks or frequency-cover index of Dyksterhuis (see Brown, 1954) may be used.

Lambert and Dale (1964) favour the use of presence/absence measures in vegetation analysis, since they form a ' self contained logical system '. They stated that all quantitative data are truncated at zero when species become absent, i.e. it is possible to measure how much a species is present but not how much it is absent, whereas in qualitative measures of frequency the absence of a species is as important as its presence. They disagreed with Greig-Smith (1964) who considered that the important differences in stands of vegetation lie in the amount of different species rather than their presence, since stray species often occur. As predicted, they found a higher information content in the purely qualitative as opposed to quantitative elements of a heathland vegetation studied. Lambert and Dale further pointed out the much more time-consuming nature of quantitative measurement over qualitative presence-or-absence measures as well as the considerable increase in computing time and costs in analysis of the former. Hyder *et al.* (1963) also supported the use of frequency for its speed and objectivity. While cover characteristics have been used in the past for describing range site and condition, density and dispersion, as inferred from frequency can also do this.

There are several problems associated with the measurement of presence or frequency which involve (a) the size of the sample, (b) the different sizes of the plant units of the vegetation and (c) the pattern of distribution of the individuals (Kershaw, 1964). The usual method of measuring presence or frequency is by using quadrats, although point quadrats in large numbers over a prescribed area have also been used. Often a single minimal area is used simply to detail presence by listing.

Clearly, as the size of the sampling area is increased, the greater is the chance of containing a given species or set of species. Thus, it is important when comparing the results from different analyses to do so on the basis of the same sized quadrats. It is obviously advantageous to make the size such that the species of interest do not yield 100 percent frequency. As we shall see later, it is possible, by plotting the variance of the data for frequency against the sampling plot size, to identify peaks on the resulting curve, which represent scales of pattern in the data, the lower-order scales having been found to correspond with plant size. This is usually difficult or impracticable to predetermine, but it serves as an indication that a sample area less than the size of the predominant plant unit is likely to add unnecessary variability to the data, and probably a size which would usually contain on average at least two or more plant units would be preferable (see Chapter 2).

Hyder *et al.* (1963) found that techniques for determining appropriate quadrat size were not as objective as might be desired. The more complex the composition of the vegetation, the more complex the problem of sample size, since species of differing frequencies, distribution and size are not sampled equally by one size of quadrat. In this respect, Tueller *et al.* (1972) have devised a convenient variable sized quadrat. Cain and Castro (1959) pointed out that the number of species found increased as the logarithm of the sample area increased. which is the basis of the species-area curve. However, Rice (1967) suggested this to be a very inadequate concept, because of the difficulty of objectively defining the most efficient ratio. Curtis and McIntosh (1950) suggested that the quadrat size should be one to two times the mean area of the most common individual. For randomly distributed species (which is not the normal situation) sampling at these sizes would give approximately 63 and 86 percent frequencies respectively for these species. Hyder *et al.* (1965) found by trial and error that a quadrat 5 cm (2 in.) square was satisfactory for sampling blue grama, but one 40 cm (16 in.) square was required for sampling the remaining species in short grass range in Colorado. Thus, they suggested using two quadrat sizes in each sampling. This is not difficult to do since the smaller quadrat can be part of the larger one.

Unless the objective of the study is clearly stated beforehand, it is sometimes difficult to know which are the species of importance to the analysis. Sampling to determine changes in frequency of, say, the dominant species blue grama is relatively simple, but in lesser known grasslands the important species or group of species may not be sufficiently known. For this reason, particularly to ensure that the information content of the sample is kept large, Williams (1971) suggested the use of a generously sized quadrat and found the classic meter square suitable for most applications.

When there are large differences in size between plant units, the sample size will have a considerable effect on measured frequency (Kershaw, 1964). This is particularly so when shoot frequency is used. Shoot frequency is more likely to be the form of frequency measured because it is uncomplicated by the need to ascertain if the portion is rooted or not, particularly since frequency is used for its simplicity and rapidity. In practice, a size of sampling unit which compromises between avoiding 100 percent frequency in the most prominent species and gross underestimation of the very small plants should be determined by experience.

Usually, it has been found that plants are rarely distributed at random in a population, but show more or less aggregated distribution. Both Kershaw (1964) and Greig-Smith (1964) have amply demonstrated this aspect in relation to sample size. It is for this reason, almost more than any other that the sample size should be stated in relating frequency data. Brown (1954) and Greig-Smith (1964) both reviewed this problem in relating frequency data to abundance or density estimates. Blackman (1935) described a mathematical relationship between density and percentage absence for a randomly distributed population of individuals, but this relationship is seldom applicable. However, as discussed later, it can be applied to micro-scale methods, where botanical composition is being estimated from the microscopic examination of macerated samples resulting from excreted or digested material (Free, Hansen & Sims, 1970). In this case the plant fragments are dispersed at random in the field of the sample (Fig. 4).

While the measurement of frequency is strongly affected by the patterns of distribution of individuals, conversely, frequency data can be used to measure this pattern. Not only can the larger scales of pattern associated with floristic and vegetational differences be described, but the intra-population scale of pattern associated with clumping or contagious distribution and other forms of random or non-random distributions also.

There are some other derived values which are related to presence. They are used to describe different manifestations of presence (e.g. fidelity), which unqualified presence sampling overlooks, or they take into account some additional property of the vegetation such as structure (e.g. dominance).

(*e*) *Derived values*

Fidelity has been defined by Cain and Castro (1959) as a measure of the relative exclusiveness of occurrence of a species in a distinct plant association or group of related associations (alliance). They considered there are two concepts embodied in this term, one is the degree of exclusiveness and the other is the relative success of a species, as measured by vigour or number in one particular community out of several in which it occurs. Fidelity is an important concept in the Braun-Blanquet system of plant sociological analysis (Cain & Castro, 1959; Greig-Smith, 1964; Shimwell, 1971). The five fidelity classes used are : (i) Strange species—rare or accidental; (ii) Indifferent species—companion species; (iii) Preferential species; (iv) Selective species; and (v) Exclusive species, which together are the characteristic species. Braun-Blanquet considered the characteristic species as primarily decisive for the floristic individuality of a community and collectively as the best indicators of the ecological condition of the community; they permit an estimate of the stage of development attained by the community; they allow conclusions on the present and former distribution of certain communities and their natural affinities, and thus their classification on a floristic basis. However, the assignment of a species to a fidelity class is very subjective and depends greatly on the experience of the investigator as well as on the amount of information available.

Doing (1972) proposed improvements to this concept, firstly in expanding the scale to a 10-point one and secondly in providing clearer definitions for these scale categories. A more objective definition of fidelity value is available from the more recent approach of Dale and Anderson (1973).

Constancy is a measure of the regularity of occurrence of a species in a set of stands. Thus, it differs from frequency which measures the homogeneity of dispersion of individuals of a species within a stand or the combined frequencies from a set of stands (Cain & Castro, 1959). Constancy has been used only in the phytosociological method of vegetation analysis of Braun-Blanquet (Greig-Smith, 1964; Shimwell, 1971).

Dominance is a characteristic which describes the importance of one or a group of species over all the others on the basis of one or more of the properties of the vegetation. The concept is most easily applied when there is a distinct life-form difference within the vegetation which has some obvious and sensible meaning. For example, in a woodland vegetation trees are clearly dominant over herbs, which is simply a physiognomic-structural dominance of form. It may have some real meaning in terms of the relative size, weight or volume of the one form over the other, as suggested by Curtis and McIntosh (1950) and Daubenmire (1968); but not necessarily, because these attributes may not be measured readily. The definition is usually a subjective eye-determination based on the appearance of the vegetation. This is simple dominance of form, but in such multi-layered vegetation a further dimension is included which describes floristic dominance within one or more of the strata. Again, there is no one attribute which necessarily defines dominance, for it may be on the basis of either number, weight, or area, according to the appropriateness of any one of these attributes to the situation. Cain and Castro (1959) have detailed a number of different categories of dominance which they called : (i) ecological-(physiological) dominance, (ii) physiognomic-dominance and (iii) family-dominance. With ecological-dominance, they considered area of coverage to be the attribute most closely expressing dominance, since this is the best approximation to volume or the relative space occupied by species. In the case of general ecological studies, the foliage cover in each stratum is used, in forestry either crown class or basal area, in grasslands basal area. Physiognomic-dominance relates more usually to life-form-dominance rather than individual species-dominance. It may include Oosting's (1956) aspect-dominance caused by the scattered occurrence of a distinctive species, or the flowing and apparent but ephemeral-dominance of a species. Family-dominance may be useful where members of species and genera of a distinctive form comprises a dominance group, e.g. grass-dominance, legume-dominance, etc.

Numata (1966) has considered the various methods of estimating dominance and pointed out the various levels of expressing it as one, two, three or four-factor dominance. The first of these is the use of cover, weight, frequency or number separately as the defining criteria of dominance. An example of this is De Vries' (1937) use of dominance frequency or dominance percent (the percent of sampling units in which a species ranks first in weight). Subsequently (De Vries & Ennik, 1953; Dirven, Hoogers & De Vries, 1969) the dominance percent has been used widely in the Netherlands in pasture studies and in the agronomic classification of Dutch grasslands.

Two-factor dominance has been used by Numata (1966) as the summed dominance ratio (SDR), or the summation of cover and height, which correlates well with weight and has been found very applicable to grassland surveys. The summation of basal area ratio and density ratio has been found more applicable to forest vegetation. Three-factor dominance has been used by Cottam (1949) as his DFD index which is based on dominance (as cover), frequency and density. Four-factor dominance based on density, basal area, size class and frequency has been used by Weaver and Clements (1938) as the phytograph method.

(f) Agronomic indices

The agronomic value of grasslands depends on yield, botanical composition and agronomic value of individual components. Various attempts have been made to combine these attributes in some form of agronomic index of grassland. These indices allow for agronomic comparisons of grasslands of different botanical composition. 't Hart and De Vries (1949) proposed an 'Agronomic Value' (gQ or grade of quality) calculated from the composition of the sward and an agronomic evaluation number for each species of grass and legume (De Vries & De Boer, 1959). Daget and Poissonet (1972) have more recently developed an index termed 'Pastoral Value' which is derived from the relative species frequency, the contribution of frequency (frequency of a species as a percent of the sum of all species) and a specific quality factor (an empirical palatability factor).

The condition and trend classification developed in North America to evaluate rangelands have further extended this approach by considering various environmental and land use criteria, though still retaining floristic composition as the central attribute in the scheme. Stoddart and Smith (1955) have outlined the approach, clearly enunciating that the basis for range condition classification rests upon the theory that vegetation is the product of the environment, a cause-and-effect relationship. While our views today are less doctrinal and, hopefully, more enlightened, Poulton and Tisdale (1961) have shown how vegetation and soil factors can be combined quantitatively in a single integrated classification.

5. Sampling strategy

It is most unusual, and indeed generally impossible, to measure the vegetation *in toto* and so some sampling procedure is used. The theory behind this process is discussed in Chapter 2 and also by Jolly (in Brown, 1954; Greig-Smith, 1964; Kershaw, 1964). Any sampling is an attempted approximation of the real situation and it is obviously the endeavour of the investigator to approximate the real situation to within certain stated limits.

(a) Level of accuracy

A sampling strategy takes cognisance of a considerable number of conditions and the first and foremost of these is the level of accuracy desired. It is possible, by taking a set of samples which is obviously more than the number necessary for an adequate sample, to calculate the number of samples needed to estimate within certain limits of the population mean. Greig-Smith (1964) outlined a useful general rule whereby a set of samples of, say, 5 is taken and a mean value calculated. The number of samples is then increased to 10, 15 and so on and their mean values plotted against the sample number. When the fluctuations of the resultant graph become minimal, an adequate sample number has been reached.

Walker (1970) pointed out that since frequency data are distributed binomially, the number of quadrats directly determines the confidence limits of the values obtained. In his studies on a Rhodesian perennial grassland he found that a value of 50 percent frequency from 100 quadrats had a confidence interval of 40-60 at the 95 percent probability level. Therefore he took 150 quadrats from each site.

The level of accuracy obviously depends on the scale of the study and the level of detail which is being investigated. Indeed it may be necessary to define the level of detail by the level of sampling which is physically possible under the particular circumstances. There are ways that the possible limits of sampling capability can be extended, as we have discussed earlier. Aerial photography can aid in extrapolating conclusions, so that an area can be stratified and only certain parts of it sampled. Considerable efforts are being expended on the use of low-level or large-scale aerial photography for this purpose (Carneggie, Wilcox & Hacker, 1971; Driscoll, 1971; McCown, Tolson & Clay, 1973).

(*b*) *Size of sampling unit*

The size of the sampling unit considerably affects the number of samples taken, for there is a strong inverse relationship between sample size and sample number. However, as McIntyre (Chapter 2) has pointed out, the relationship favours number, i.e. it is always more efficient to have larger numbers of small sampling units than fewer large ones, even though the total area sampled may be the same.

The dispersion and form of the vegetation interacts with sample size as well. In vegetation where the plant individuals are sparsely distributed, a small sample size may be an inefficient measure of composition and if the individuals vary considerably in size and form themselves (e.g. bunch grasses, creeping grasses or annuals) considerable bias for or against the different components may result. Furthermore, the particular values being measured may be strongly influenced by the sample size, as is the case with frequency. Both Greig-Smith (1964) and Kershaw (1964) have discussed these problems in detail.

Sometimes in large-scale sampling of natural vegetation, as with mapping, it may be possible to take only one sample from each vegetation type. In this case it is necessary to know whether the sample adequately represents that vegetation unit. This may be done by constructing a minimal area curve which is simply a plot of cumulative species number against increasing area of plot. Usually, this is done by successively doubling the plot size. The resultant curve is characteristically sigmoid, shallower in vegetation of few species and deeper in species-rich ones. The mean point of inflection indicates the minimal or representative area. A detailed discussion of this concept has been given by Cain and Castro (1959), Greig-Smith (1964) and Kershaw (1964). Hyder *et al.* (1963) found the point of contact of the 45° tangent to the species area curve approximated the optimal quadrat sampling size rule, which Curtis and McIntosh (1950) suggested as being between one and two times the mean area of individuals of the most common species.

Kershaw (1964) also showed how the mean square for species density or cover fluctuates according to various scales of pattern when plotted against increasing plot sizes. The peaks in the curve denote optimal scales of patterns increasingly for growth form of the dominance components, dispersal range, topographic variability and so on. Such an analysis can be used to ascertain in a preliminary way first of all what size samples should be used to contain the first level or, if a regular pattern of sampling is adopted, what sample spacing should be used to avoid harmonic bias with topographic pattern.

(*c*) *Sample number*

As has already been stated, sample size and sample number are to a large degree alternatives, although the relationship favours number. However, Williams (1971) has warned against the sample size becoming too small, particularly with frequency analysis, because of the reduction in information contained by the sample. He gave the extreme case in rain forest studies where the size of sample is governed by the information it contains rather than relying on a fixed size where the information content varies a great deal.

The number of samples required is obviously different for the more abundant species compared with less common ones and Greig-Smith (1964) has suggested that sampling can be left off at a certain level for the more common species while continuing for the lesser ones. The sampling strategy adopted also has an effect on the number of samples taken. Thus, with recurrent sampling systems, more samples are required when they are distributed randomly than when a systematic layout is used and succeeding samples are taken from the same positions on subsequent occasions. The theory for this has been outlined by McIntyre (Chapter 2).

In practice, the number of samples taken will be determined by the experience of the investigator in consultation with his analytical advisor, after consideration of the importance of the information being sought, the degree of precision required, and the physical limits of the study. Greig-Smith (1964) has given a rule of thumb means of checking the adequacy of the numbers of samples being taken which was outlined in the discussion on levels of accuracy.

(*d*) *Distribution*

Plants in a vegetation may be distributed randomly, regularly or contiguously.

Actually, the first two distributions are rare while the last, though the usual situation, varies greatly. This obviously has some effect on the sampling strategy, particularly in relation to the distribution of samples, which may themselves be random, systematic or stratified random (Chapter 2). Williams (1971) maintained that, since much of the work on vegetation is concerned with pattern, a rectangular grid, preferably with equal spacing in each direction, is necessary. Only in studies of distribution of single species, where an estimate of density with its fiducial limits is required, should random placement of samples be used. Williams has further objected to random sampling on the basis of the irregularities and inconsistencies of scale to which no meaning can be assigned.

Cain and Castro (1959) have pointed out that irregular sampling, as with the method of so-called ' random throws ', is definitely not random sampling and may be subject to considerable bias. Random sampling proper can only be on the basis of locating each point by two randomly selected coordinates, which is a slow procedure. Some reduction in the difficulties of placing random samples in a large area can be achieved by dividing the area into strata on the basis of some clearly defined criteria such as aspect, altitude, topography, soil or moisture regime. This is usually the case in locating plots for yield estimates. On the other hand, Williams objected to the practice of ' subjective sampling ' where samples are taken from ' typical ' or ' representative ' sites. Not only is there the obvious problem of bias, but also transition zones are not sampled. However, some form of selective grid analysis would seem to be applicable here, particularly with the availability of various forms of remote sensing. In relation to this, Switzer (1971) has proposed a method in which vegetation patterns can be ' grown ' or simulated if the main characters and restraints are known.

The measurement of plant distribution may, in fact, be the objective of some studies so that the deficiencies of certain measuring techniques in measuring composition on the one hand may actually be used as a measure of the form of distribution on the other. The fact that different quadrat sizes may give different estimates of the type of distribution (Kershaw 1964) can be used to measure pattern as with the mean square-plot size technique (Greig-Smith 1964, Kershaw 1964). Pielou (1959) used the variable point-to-plant distances to generate her Index of Non Randomness.

(e) Pattern recognition

Pattern recognition is the subject of the chapter by Dale (Chapter 8). As Williams (1971) has indicated, modern classificatory methods rest more or less on associations between the attributes recorded (usually species). The data are examined to establish which species almost always or almost never occur together and from this groupings are established. Thus sampling units must include enough information for the associations to be made. On this basis lies the justification for using the information content of the sample as the criterion for sample size, rather than the conventional area restriction, as in rain forests with high floristic diversity but low floristic frequency (Williams, 1971).

Empirical methods of pattern recognition were embodied in the Braun-Blanquet (see Cain & Castro, 1959) system of vegetation study and it is interesting to note that Dale (Chapter 8) found that numerical methods of combined ordination and classification are conceptually rather similar. Early methods of quantitatively expressing similarity groups came through the use of contingency tables and χ^2 analysis (Greig-Smith, 1964). However, there is no substitute for the necessary interpretation by the ecologist or agronomist and, as Williams (1971) has said, both the analyst and ecologist are needed together, each with some insight into the other's domain.

(*f*) *Type of sample*

Although there are many types of samples, they can be considered as finite, i.e. samples which have a definable area (quadrats) or point location (point quadrats), or non finite, i.e. samples defined by plant density considerations and therefore of no fixed size (plotless samples).

(i) *Quadrats* are probably the most used form of sample, and they vary considerably in size and shape. Sample size and number have already been discussed. By definition a quadrat is square, but various shapes from rectangular to circular are now referred to as quadrats. Both the size and the shape can have an effect on the sampling efficiency. Rectangular quadrats can be managed more easily without having to lean over or trample them and this, Kershaw (1964) considered, may account for their greater accuracy. In yield studies McIntyre (Chapter 2) reflected that rectangular quadrats are more accurate because they include more of the small point-to-point variability and therefore less between-sample variability.

There is an edge effect associated with quadrats which becomes proportionately larger as the size of the quadrat decreases or becomes more rectangular, and smaller as the quadrat size increases or becomes circular. The edge effect arises from the often experienced difficulty of deciding whether a plant is in or out, because it may or may not be rooted within the quadrat, what to do with lodged vegetation and the disturbance caused by the placing of the quadrat in the first place. In this latter situation, particularly in long vegetation, open-ended quadrats are useful. The discussion of sample size is pertinent here to quadrat size.

Quadrat sampling can be used for measuring all the criteria of composition, although, in fact, this type of sample is less suited to area measurements than to others. Quadrats are usually placed independently, according to the accepted sampling strategy, either regularly gridded or along predetermined transects, in a stratified random or completely random arrangement. A belt transect may be considered as a set of contiguous quadrats arranged in one line.

(ii) *Point samples.* These are usually referred to as point quadrats but they are a sample, theoretically, without area. It follows to the ultimate level the advantages of reducing sample size while increasing sample number. In this process, the within-sample variation is reduced to zero, so that all the variation is that related to between-samples. This form of sampling has been widely used since first developed by Levy and Madden (1933) as the following selected bibliography illustrates (Brown, 1954; Cain & Castro, 1959; Greig-Smith, 1964; Kershaw, 1964; Poissonet & Poissonet, 1969; Anon, 1970; Heslehurst, 1971; Becker and Crockett, 1973; Poissonet *et al.*, 1973). There are, of course, problems associated with approximating such a sample, since it is physically almost impossible to achieve. The usual approximation is that of a needle point, but even this has a finite area, the implications of which have been discussed by Goodall (1952) and Warren-Wilson (1963). Cover is overestimated increasingly as the size of the point increases. Conversely, the finer the point becomes the more difficult it is to maintain rigidity and alignment of the point apparatus, and this imparts errors and often bias to the measurements. This problem has provided a fruitful field for innovation (Brown, 1954; Long *et al.*, 1972; Owensby, 1973).

Radcliffe and Mountier (1964a) have discussed the various types of information that can be obtained from point sampling and what aspects of pasture composition this expresses. These are (i) the first plant hit, (ii) crown hits, (iii) first hit on each species (' cover hits ') and (iv) all hits on each species (' total hits '). The first plant hit is a rapid and easy measure of the dominant species in situations of fairly uniform growth-form. Where the growth-form of the major components is quite different, as with most grass-legume situations, the estimates are biased strongly in favour of the legume component. Crown hits are practicable only in certain plant associations where the plant units are discrete and generally layered. The first hit on each species gives an estimate of the area of ground covered by each species, but does not indicate the extent of the layering of the vegetation. Goodall's (1952) ' percentage cover ' is the first hits of all species per 100 points. Total hits, or all hits on each species, provide a measure of bulk (vertical and horizontal density) when expressed as total hits per 100 points. Sward composition can be expressed as the percent total hits on a species relative to the total hits on all species. ' Cover repetition ', which is a measure of the layering of the vegetation, may be estimated from the relationship of all hits on a species to the number of first hits on that species. Finally, a measure of sociability can be made by considering the total number of needles that hit a particular species and separating them into those that have hit additional species with the same needle and those that have not.

A method for overcoming the problem of point size was proposed by Winkworth and Goodall (1962), in which a crosswire mounted in a sighting tube was used to define the point. This overcame the mechanical disadvantages of the pin and its finite area, but it was difficult to use in anything but short vegetation and almost impossible to use if all hits throughout the canopy were desired. Ibrahim (1971) has used the technique in desert grassland, while Burzlaff (1966) has used a telescope mounted on a tripod which describes a circular transect. It is also used in micro-scale analysis using a microscope (Stewart, 1967), in which randomly located points were marked on the eyepiece.

The loop method of Parker and Harris (1959) is essentially a compromise between the problems encountered in accurately using a pinpoint to measure cover for evaluating range condition and the easier, more rapid and objective measure of frequency from an area sample. The bias imparted by the loop size in the measurement of cover was in most cases insufficient to obscure the rather large differences being measured. In more detailed studies its use has been criticised and its accuracy has fallen short of other methods (Johnston, 1957). Hutchings and Holmgren (1959) have suggested ways of overcoming bias by applying correction factors to the loop frequency data in measuring plant cover, but Francis, Driscoll and Reppert (1972) have found the bias to be unacceptable and impossible of correction. Reppert and Francis (1973) have subsequently proposed new guidelines for interpreting data obtained by the 3-step method for evaluating range condition (Parker & Harris, 1959).

A comprehensive and recent review of the point quadrat method and its application to a wide range of situations has been made by Heslehurst (1971) and will not be repeated again here in the same detail. Apart from the original use of the method in measuring the floristic composition of vegetation in terms of cover, the method has been adopted for measuring various structural features of the vegetation. These are largely related to studies of growth and photosynthesis, the measurement of leaf area being important in studies of photosynthetic efficiency of different vegetation types, and variation and dispersion of the foliage important for the relative efficiency of light utilization by the herbage. Changes in leaf area are brought about in the course of pasture establishment by seasonal interplay between the pasture components resulting from grazing, fertilizing, the composition relative to the initial seed mixture, or subsequent incursions of adventive plants into the system. This is a rather more specialised application of the method which has been largely developed in the detailed studies of Warren-Wilson (1965) and Heslehurst and Wilson (1971) (see also Heslehurst, 1971) of vegetation canopy profiles, calculation of light extinction coefficients and irradiance characteristics.

Spedding and Large (1957) have developed a point quadrat method for the determination of density at vertical intervals by recording hits along graduated pins. Point quadrats can be applied effectively to sampling cut herbage and, as already discussed, to microscopic examination of faecal or digesta material.

The point quadrat relies on the advantage gained in estimating variability by reducing the sample size while increasing sample number. However, what is often overlooked is the fact that the information content of each sample is very much reduced; actually, in a great many samples it is zero. Philip (1966) suggested from his analysis of the use of point quadrats that the accuracy is related more to the number of contacts than to the number of points, which accords with Williams' (1971) statement that it is the total number of records which is important.

The number of points to be taken is obviously influenced by many factors besides that of level of accuracy desired. Type of attribute—frequency or cover and whether it is canopy, profile or basal area cover; type of vegetation—floristic richness, openness and so on are also important. However, Goodall (1952) has suggested that the variance between two points placed at random within parts of a larger area can give a rough indication of the number required. Warren-Wilson (1965) has further discussed this in relation to inclined point quadrats, a modification he developed to study interception of light by foliage.

Although the initial use of point quadrats was based on single point locations (Cockayne, 1926), the application since then has usually been with frames of points, usually ten in linear arrangement. This means that unless the frames are placed contiguously along a single line or transect, the points are aggregated by tens. Kemp and Kemp (1956) have investigated this situation and found that, while it was desirable to reduce the number of points per frame, on theoretical grounds this disadvantage was usually compensated for by the greater number of points which could be sampled in the same time with multiple point frames. If it is desired to have some estimate of small-scale patchiness, a frame with at least two points is necessary. Conversely, with patchiness it may be creating bias, as Goodall (1952) suggested, since the probability of hitting a species patch is less between frame positions than within.

Whether points are taken at random, stratified random or at systematic positions is essentially a consideration similar to that with quadrats. Point quadrats are used more often as an estimate of cover on a particular piece of vegetation which has been defined by an agronomic treatment or by some means of pre-selection. Although now used less as a survey technique, Shockey and De Coursey (1969) have used a systematic point sampling procedure on a large-scale land use survey and considered systematic sampling to be more than three times as efficient as random sampling. Sampling with a wheel-point apparatus (Tidmarsh & Havenga, 1955; Roberts, 1972) is a systematic point-sampling procedure, as also is the step-point method of Evans and Love (1957) in the California Soil Survey (Wieslander & Storie, 1952).

(iii) *Plotless sampling*. This method of sampling was originally developed for use in forests where the individual plant units are distinct. Theoretically, the density of a stand of individuals is inversely proportional to the distance between the individuals. Cottam and Curtis (1949) and Curtis and McIntosh (1950) proposed and used a workable procedure based on distance measures between randomly selected pairs of trees along a transect. Later Cottam and Curtis

(1956) developed the point-centred quarter method, which proved to be the least variable of four distance measures then in use. Cottam, Curtis and Hale (1953) suggested that a reasonable level of sampling had been achieved when the species of primary interest had occurred more than 30 times. They also warned that random populations tend to give rather variable results with this method. Pielou (1959) has used point-to-plant distances in determining the distribution, random, regular or aggregated, of plants. Provided an independent estimate of density is available, the distribution can be calculated by the use of Pielou's Index of Non-Randomness. This method has overcome the problem of obtaining clumping, which is the product of fixed sample size when using quadrats.

Dix (1961) applied the point-centred quarter method to sampling a perennial bunch grassland vegetation in Western North Dakota. Important values (relative frequency + relative density) were calculated from the measured distances to four nearest shoots in the quarters. The method appeared more rapid than line intercept or point frame measures. It appeared to be very efficient in detecting slight differences between stands or changes within stands. Penfound (1963) has suggested a modification of the method of Dix, which essentially only provides data on frequency and density, by incorporating a measure of weight. In this case, the shoot forming the nearest neighbour in each quarter is harvested and its dry weight determined. He pointed out the problem that the quarter method tends to magnify the role of small species and minimises that of large ones and this must be taken into consideration. Risser and Zedler (1968) found the method also decreased the relative density of aggregated species and increased that of less aggregated species. Good and Good (1971) found the method underestimated density in prairie vegetation.

In rainforest sampling, Williams and Webb (1969) have extended the plotless sampling technique to that area defined by the 20 nearest trees to the randomly or systematically selected reference point.

(iv) *Core sampling.* De Vries (1937) developed a method of sampling on short grazed pastures in the Netherlands in which a 25 cm^2 core of turf was taken for dissection and analysis in the laboratory. From this he was able to determine species frequency and the rank order in terms of dry weight that each species occupied. This has remained the main method of sampling in the Netherlands (Dirven, Hoogers & De Vries, 1969). Hutchinson (1967) and Hutchinson, McLean and Hamilton (1972) have used the technique on closely grazed sheep-swards, where conventional herbage clipping techniques have been found to be unsatisfactory. Using the apparatus Mitchell and Glenday (1958) developed for tiller sampling in dense swards, it was possible to measure both aerial and underground plant material. The method has been extended by using the cores as a portable set of pasture standards against which field samples can be matched and scored (see Chapter 4).

(v) *Charting and photographic methods.* Brown (1954) has discussed the use of charting methods, such as the pantograph charter or plotting onto squared paper. These methods are not used very much now, since they are slow and very tedious, their application being largely related to life-history studies of individual plants. Photographic methods have acquired wider acceptance because they not only provide a permanent exact record of the sample areas but they can be taken rapidly in either two or three dimensions (Wimbush, Barrow & Costin, 1967; Pierce & Eddleman, 1970; Wells, 1971; Goodwin & Walker, 1972) (Plate 5).

Both charting and photographic methods have limitations due to the height and closeness of the vegetation. They are two-stage methods which require further measurement to achieve quantitative data. If the species are readily distinguishable, this may fairly easily be accomplished by superimposing a grid over the photo or chart. With photographic transparencies a grid may be projected over the top of the image. Usually the photographing or charting is confined to a particular time of the year so that the growth stage of the plants

PLATE 5. Camera stand for recording tall vegetation (Goodwin & Walker, 1972)

is fairly comparable from year to year and at a level which allows adequate resolution.

(*g*) *Comparison of methods*

There is a considerable number of published accounts of studies comparing different methods of measuring herbaceous vegetation. Brown (1954) has reviewed the earlier ones, but more recent studies include those of Mountier and Radcliffe (1964), Johnston (1957), Lyon (1968), Dirven, Hoogers & De Vries (1969), Walker (1970), Poissonet *et al.* (1973). From these, it is clear that no one method is a panacea for all situations, but that the choice of a method and its implementation is a very individual matter. In general, it appears that in pasture studies where the floristic composition is fairly simple, the vegetation fairly short, and ground cover mostly complete, point quadrat sampling for cover, or harvesting or ranking for yield are the favoured methods. Thus Radcliffe and Mountier (1964a) concluded that the point quadrat method was clearly suited to the measurement of area rather than volume and that cover hits (first hit on each species) and total hits (all hits on each species) were the clearest expression of species composition. They (1964b) also pointed to the problem that various vegetation heights pose. Daget and Poissonet (1971) and Poissonet *et al.* (1973) have proposed various ways of handling this situation by using point quadrats or line intercepts according to the height of vegetation. In recognising the need for rapid and simple sampling techniques and the need to use different techniques for different situations, they have attempted to evaluate precisely several very different methods so that some means of relating the results can be achieved. The rapidity of frequency sampling combined with the simplicity of point analysis and the ease of ranking for dry weight have been examined to develop relations between frequency (expressed as ' contribution of frequency '), cover (expressed as ' contribution of cover ') and relative biomass.

In areas where the vegetation is more open, more floristically diverse and the studies necessarily more extensive, Hyder *et al.* (1965, 1966) and Walker (1970) considered presence/absence or frequency sampling to be generally the more appropriate of the various methods investigated.

Most studies are likely to combine several measurements in one, with presence/absence likely to be the central one, because of the speed and objectivity of its measurement. Accompanying measurements may be either qualitative or quantitative. However, a clear concept of how the data are to be used must first be formulated, since as the complexity of the data increases, so does the analysis. While this poses less of a problem now than previously, it does mean that the analysis must play a more significant role than previously.

III. REMOTE SENSING

In recent years there has been a significant development in the use of sensors for measuring various environmental parameters. Aerial photography in the traditional sense may now be regarded as just one form of remote sensing, but it plays a central role in many of the developments in remote sensing because of its suitability for recording the information provided by many of the sensors. Significant progress has been made since the initiation of the space exploration programme and a great deal of work is currently centred around this. However, sensing can also be carried out from high-altitude balloons (Girard-Ganneau & Girard, 1974) as well as from the ground.

Remote sensing is essentially a process of indirect or secondary measurement of environmental (vegetation) parameters. It concerns the detection and recording of spectral information. Conventional aerial photography is the most widely used application of spectral recording and, because of the accuracy of the images which can be obtained, perhaps the most versatile. However, the photographic wavelengths are only a small segment of the total electromagnetic spectrum, which ranges from cosmic rays at wavelengths of $<10^{-12}$ cm to Hertzian waves at $>10^4$ cm (Baumgardner, 1970). At present most remote sensing is confined to optical wavelengths of approximately $0{\cdot}3$ to 15μ, which encompasses about three times the photographic band (Hoffner & Johannsen, 1969; Baumgardner, Leamer & Shay, 1970). However, some sensing is also being carried out in the Hertzian band by means of radar (Holter, 1970; Attema *et al.*, 1974).

1. Optical sensing

Within the bandwidth of optical sensing Hoffer and Johannsen (1969) have indicated two important regions :

(i) the reflective region of $0{\cdot}3\mu$ to 3μ.
(ii) the emissive region of 3μ to 15μ.

They are not mutually exclusive, however, since as reflectance falls off towards the 3μ wavelength so emission increases.

The reflective region may be further divided into the visible and non-visible areas. The visible area is that in which photographic recording may be used, while in the non-visible area some form of optical-mechanical scanning must be applied. The spectral reflectance of vegetation is primarily determined in the visible portion of the spectrum by the chlorophyll pigments and in the reflective infrared by water. Thus, information on reflectance can provide useful information on important structural and physiological properties of vegetation or plants (see Gates, 1970).

The emissive region concerns thermal radiation, patterns and gradients of which are being studied in relation to time sequences (Johnson, 1969). At normal ambient temperatures the spectral composition of emitted radiation from plants is entirely in the infrared region with a peak at about 10μ. Since this is outside the visible range it may be measured by infrared radiometer (Gates, 1969).

As improved sensors become available, increasing use can be made of non-photographic methods of sensing. However, it is true to say that photographic methods will remain as the basis of much sensing. Multispectral scanning, both as imagery and digital output, has been given great impetus since the launching of the ERTS satellites and much of the current work is based on this information. For example, Schubert and MacLeod (1974) have found that numerical classification of digital reflectance data from ERTS-1 gives a tolerable physiognomic vegetation classification at a mapping scale of 1 : 20,000. They argued that the IBP classes of vegetation based on density, structure and seasonal persistance are all related to optical characters of plant communities which are largely described by reflectance values. So far, resolution of vegetation from space has not been at much less than the physiognomic level.

There are two problems associated with remote sensing of vegetation. The first relates to photographic methods and the difficulty of objectively analysing imagery. There are some developments in the field of computer picture processing which are as yet at an early stage (Rosenfeld, 1968). Much imagery is not required to be objectively analysed however. The second problem relates to the sheer weight of data that may become available from a sensing source. Not only is it necessary to be able to handle such large amounts of data, but it is also necessary to set up relationships with vegetation characteristics. This requires a considerable amount of ground truth (de Boer *et al.,* 1974) and it is very easy to outstrip this capacity. As with any sets of measurements, sampling and statistical problems exist (Kelly, 1970).

2. The application of remote sensing to vegetation measurement

There are obvious applications of remote sensing for the indirect measurement of vegetation at different levels which a number of recent publications have outlined (Stewart, 1968; Johnson, 1969; NASA, 1970; Baumgardner, 1970; National Research Council, 1970). It appears likely that the data from remote sensing will be applicable to an even wider number of vegetation properties than hitherto. For example, since perception of radiation concerns energy, there are likely to be important applications to the functional aspects of vegetation as well as the more obvious ones associated with physiognomy, structure and composition. The Proceedings of the International Symposia on Remote Sensing of the Environment, Ann Arbor, Michigan, which take place every two years provide a multitude of applications of remote sensing, as also does the journal *Photogrammetric Engineering.* As outlined by several of the above publications, the applications in crop logging, production prediction, disease epidemic monitoring and so on are considerable.

IV. PRACTICAL CONSIDERATIONS

At first sight there seems to be a formidable array of techniques for measuring vegetation and anyone setting up to do this for the first time is likely to be justifiably confused. However, as Brown (1954) has said, all these methods measure one or more of only four basic values of the vegetation. In choosing a method it must first be decided what the *sample* will be, i.e. whether the approach is one of survey and the recognition of pattern, in which case a rectangular grid with equal spacing in two directions is ideal, or whether the study is concerned with exact distributions of individual species (random, clustered, regular, etc.) with a statement of their fiducial limits, which requires that the samples be set out at random. Then it must be decided which one of or which combination of these attributes best provides an adequate measure of the vegetation for the purpose of the study, i.e. the most appropriate property to measure and the values or terms in which it will be measured. Finally, the feasibility in terms of time, manpower and sampling adequacy, must be considered and then a method of carrying out the measurements chosen or formulated. The methods available for processing and analysing the data also play a part. Particularly since the advent of the computer, much intractable data can now be handled with ease. All too often the method used has been chosen for no other reason than the fact that it was known to the investigators. Thus such methods as point quadrats, loop, etc. enjoyed a considerable following, but were not necessarily the most suitable.

The following summarizes the features of the four attributes of vegetation and attempts to give a thumb-nail sketch of the considerations to be borne in mind in formulating a programme of study.

Number is an objective quantitative measure usually applied to studies of germination and establishment or to the behaviour or response of certain species, as in weed invasions, weed control, disease effects and so on. The main problem with number as an attribute for measurement is to formulate an adequate definition of an individual, and how meaningful that definition is where very contrasting plant forms exist within the same plant community. It is also important to know in advance something about the possible changes in plant forms, since while they may be amenable to counting initially, this may not be so later on.

Cover or area has probably been the most widely measured attribute in intensive studies of botanical composition in herbaceous vegetations. This is perhaps partly due to the popularity of the point quadrat method of measurement, which is by far the best method of measuring cover. It is an ideal measure for determining the composition of swards, since the investigator is usually more interested in the relative proportion of the different leaf components present than the species individuals (in closed or dense swards individuals do not matter). However, in estimating individual species composition, basal area provides a suitable measure if an adequate definition of an individual can be made. In herbaceous vegetation, seasonal growth patterns have a marked effect on the amount of total cover, which may itself be a crude measure of bulk, though this can be improved by including height measurements and information on secondary cover strata. The lack of cover may be the point of interest, as bare ground often reflects climatic or grazing effects and is important in hydrological studies of water interception and runoff.

Weight. Composition by weight is often part of the measurement of total dry matter production in pasture studies. Traditionally, hand separation is employed and this is not so laborious if only two or three species components are involved. In more complex systems measurement may be carried out on the cut sample suitably displayed, or its composition may be estimated.

Recently, a new method using ranked estimates has been developed (the dry-weight-rank method of 't Mannetje & Haydock, 1963), which has fairly wide application. There are two problems with its use. The first is that the method cannot be used if the contribution of one component exceeds 70 percent of the total dry matter. Some users have overcome this by giving first and second rank to a species which contributes more than, say, 75 or 85 percent to the total dry matter in a quadrat. Another means of avoiding the problem is by partitioning the component in question into dry and green. The other problem is that the method will give false results if one species is always dominant in high yielding quadrats and another always in low yielding ones. This can be overcome by including an estimate of yield for each quadrat, as in the method of Haydock and Shaw (1975), using the yield score as a correction factor for the contribution of each species to the overall botanical composition (Jones, 1974). However, these problems are relatively uncommon for the majority of pastures (see Tothill, Hargreaves and Jones (1978) and Hargreaves and Kerr (1978)).

Presence. The measurement of presence/absence, frequency or occurrence is fast and objective. It is a somewhat less sensitive value for measuring botanical composition than some others, but is particularly suited to detecting pattern in vegetation and temporal changes in floristic composition. Because of the effect that sample size has on the frequency value it is often difficult to relate information from one analysis to another unless the size and number of the sampling units is the same. Presence or frequency can be measured easily and with little extra effort in conjunction with other values, such as weight, or it can be extracted from measures of number or cover. Presence/absence records, being binary in form, are computationally advantageous, thus making it possible to analyse very large data blocks.

Finally, developments in the field of remote sensing, such as the use of various spectral bands or infra-red thermal scanning in aerial photography, are allowing accurate stratification of study areas. By selectively sampling such areas and omitting intervening ones, intensive sampling may be carried out on very large areas, providing the information lost from the intervening areas is not critical to the study. It should always be realized with this procedure that the discontinuities between groups are magnified, that undersampling may result and the transition zones may hold information important to the interpretation. It should therefore not be contemplated lightly.

The rather simplified view expressed by Williams (1971) is seemingly at variance with all the detailed work that has been published on most aspects of vegetation sampling. There are obviously two extremes: on the one hand the individual who carries out his sampling with the maximum regard for theoretical correctness may be able to cover only a small amount of ground in the time available, or on the other hand if he is obliged to cover a lot of ground in a limited time, his results may be seriously lacking in accuracy. The answer must lie in between and this can be justified by the fact that the investigator, in the absence of any previous information, often cannot predetermine what level of accuracy is required, which among the plant species present are the important ones, or what weight to attach to his own or other peoples' experience. Interaction between investigator and analyst is a vital part of any study both at the planning and resolution stages.

V. CONCLUSION

Brown (1954) concluded by suggesting that point sampling offered the most promise as an effective measure of vegetation and certainly, if the literature is any guide, this prediction was well founded. The pinpoint makes possible the exact distinction between species, even in dense swards and compared with counts or with cutting and weighing, is much more rapid, and is applicable both to large- and to small-scale studies.

Williams (1971), appraising the situation almost twenty years later, on the other hand suggested that frequency methods offer the most promise for the future. Any large-scale or survey work should never use point quadrats or line intercepts; the sampling unit must always be an area. Williams further argued that the sample area should be large in relation to the plants it contains, for if it is small it implies there is no room for any but one or a few plants and false negative associations may be developed.

Frequency data can be acquired rapidly and objectively and in this respect compare favourably with point samples. However, as Lambert and Dale (1964) have pointed out, frequency data contain more information since species absences are valuable records and, furthermore, the sampling unit contains information on several species and not just one. Of perhaps the greatest significance, at least at present, is the fact that presence/absence data are binary in the computational sense and can be handled on a far greater scale than most quantitative data. The capacity of the computer to handle very large amounts of data has lent impetus to the use of presence/absence recording. Even though many agronomic studies require the expression of botanical composition in terms of proportion by weight, this can easily be combined with presence/absence records.

Because of this it is becoming more important to consider the analytical aspects of any study. As Williams (1971) so clearly stated, there should be considerable dialogue between analyst and ecologist or agronomist, because of the increasing improbability that each will be proficient in the other's field, particularly at this time of rapid advances in data processing. There is a strong interaction between what is computationally feasible or sensible, data which are ecologically sound or pertinent, and the methods necessary to acquire this information. The whole sequence must be carefully thought through and the method used must be chosen because it is a means to an end and not an end in itself.

VI. REFERENCES

AMERICAN SOCIETY OF RANGE MANAGEMENT in joint committee with the Agricultural Board (1962). Basic problems and techniques in range research. *Publ. No.* 890, *Natl. Acad. Sci. Washington D.C.* Publ. National Research Council.

ANON, (1959). Techniques and methods of measuring understory vegetation. *U.S. For. Serv., South For. Exp. Stn.* & *Southeast For. Exp. Stn.*

ANON, (1963). Range research methods. *U.S. Dep. Agric., Misc. Publ. No.* 940.

ANON, (1970). Point quadrats and their use in sampling vegetation, 1956-70. *Annot. Bibliog. No.* G200, *Commonw. Bur. Pastures Field Crops, Hurley.*

ANON, (1972). Frequency, density and cover methods for vegetation, 1953-1972. *Annot. Bibliog. No.* 1303, *Commonw. Bur. Pastures Field Crops, Hurley.*

ATTEMA, E. P. W., den HOLLANDER, L. G., deBOER, TH. A., UENK, D., ERADUS, W. J., de LOOR, G. P., van KASTEREN, H. & van KUILENBURG, J. (1974). Radar cross sections of vegetation canopies determined by monostatic and bistatic scatterometry. *In* Proceedings of the IXth international symposium on remote sensing of environment, Vol. II Michigan; Environmental Research Institute, pp. 1457-65.

BAUMGARDNER, M. F. (Ed.) (1970). Aerospace science and agricultural development. *Am. Soc. Agron. Spec. Publ. Ser. No.* 18.

BAUMGARDNER, M. F., LEAMER, R. W. & SHAY, J. R. (1970). Remote sensing techniques used in agriculture today. *In* Aerospace science and agricultural development, Ed. M.F. Baumgardner. *Am. Soc. Agron., Spec. Publ. Ser. No.* 18. pp. 9-26.

BECKER, D. A. & CROCKETT, J. J. (1973). Evaluation of sampling techniques on tall-grass prairies. *J. Range Mgmt.* **26,** 61-5.

BIRK, E. (1972). Herbivore food preferences. Thesis B.A. (Hons.). Macquarie University, North Ryde, New South Wales, Australia.

BLACKMAN, G. E. (1935). A study of statistical methods of the distribution of species in grassland associations. *Ann. Bot.* **49,** 749-77.

BROWN, Dorothy, (1954). Methods of surveying and measuring vegetation. *Commonw. Bur. Pastures Field Crops, Hurley, Berkshire, Bull.* 42.

BURZLAFF, D. F. (1966). The focal-point technique of vegetation inventory. *J. Range Mgmt.* **19,** 222-3.

CAIN, S. A. & CASTRO, G. M. de O. (1959). Manual of vegetation analysis. New York; Harper & Bros.

CARNEGGIE, D. M., WILCOX, D. G. & HACKER, R. B. (1971). The use of large scale aerial photographs in the evaluation of Western Australian rangelands. *W. Aust. Dep. Agric. Tech. Bull. No.* 10.

CHIPPENDALE, G. M. (1962). Botanical examination of kangaroo stomach contents and cattle rumen contents. *Aust. J. Sci.* **25,** 21.

CHIPPENDALE, G. M. (1968). A study of the diet of cattle in central Australia as determined by rumen samples. *Aust. North Terr. Admin., Anim. Ind. Branch, Tech. Bull. No.* 1.
COCKAYNE, L. (1926). Tussock grassland investigation in New Zealand. *In* Aims and methods in the study of vegetation. Eds. A. G. Tansley & T. F. Chipp, London; Crown Agents for the Colonies, pp. 349-72.
COTTAM, G. (1949). The phytosociology of an oak woods in southwestern Wisconsin. *Ecology* **30,** 271-87.
COTTAM, G. & CURTIS, J. T. (1949). A method for making rapid surveys of woodlands by means of pairs of randomly selected trees. *Ecology* **30,** 101-4.
COTTAM, G. & CURTIS, J. T. (1956). The use of distance measures in phytosociological sampling. *Ecology* **37,** 451-60.
COTTAM, G., CURTIS, J. T. & HALE, B. W. (1953). Some sampling characteristics of a population of randomly dispersed individuals. *Ecology* **34,** 741-57.
CROKER, B. H. (1959). A method of estimating the botanical composition of the diet of sheep. *N.Z. J. Agric. Res.* **2,** 72-85.
CURTIS, J. T. & MCINTOSH, R. P. (1950). The interrelations of certain analytic and synthetic phytosociological characters. *Ecology* **31,** 434-55.
CURTIS, J. T. & MCINTOSH, R. P. (1951). An upland forest continuum in the prairie-forest border region of Wisconsin. *Ecology* **32,** 476-96.
DAGET, P. & POISSONET, J. (1971). Une méthode d'analyse phytologique des prairies. *Ann. Agron.* **22,** 5-41.
DAGET, P. & POISSONET, J. (1972). Un procédé d'estimation de la valeur pastorale de paturages. *Fourrages.* **49,** 31-9.
DALE, M. B. & ANDERSON, D. J. (1973). Inosculate analysis of vegetation data. *Aust. J. Bot.* **21,** 253-76.
DAUBENMIRE, R. (1968). Plant communities. New York; Harper and Row.
DAVIS, I. F. (1964). Diet selected by sheep grazing on annual pasture in southern Victoria. *Proc. Aust. Soc. Anim. Prod.* **5,** 249-50.
DE BOER, TH. A., BUNNIK, N. J. J., VAN KASTEREN, H. W. J., UENK, D., VERHOEF, W. & DE LOOR, G. P. (1974). Investigation into the spectral signature of agricultural crops during their state of growth. *In* Proceedings of the IXth. international symposium on remote sensing of environment. Vol. II, Michigan; Environmental Research Institute pp. 1441-55.
DE VRIES, D. M. (1933). De rangorde-methode. Een schattings-methode voor plantkundig graslandonderzoek met volgorde-bepaling. *Versl. Landbouwkd. Onderz.* (Agric. Res. Rep.) No. 39A, 1-24.
DE VRIES, D. M. (1937). Methods used in scientific plant sociology and in agricultural botanical grassland research. *Herb. Rev.* **5,** 187-93.
DE VRIES, D. M. & DE BOER, TH. A. (1959). Methods used in botanical grassland research in The Netherlands and their application. *Herb. Abstr.* **29,** 1-7.
DE VRIES, D. M. & ENNIK, G. C. (1953). Dominancy and dominance communities. *Acta. Bot. Neerl.* **1,** 500-5.
DIRVEN, J. G. P., HOOGERS, B. J. & DE VRIES, D. M. (1969). Interrelation between frequency, dominance and dry-weight percentages of species in grassland vegetations. *Neth. J. Agric. Sci.* **17,** 161-6.
DIX, R. L. (1961). An application of the point-centered quarter method to the sampling of grassland vegetation. *J. Range Mgmt.* **14,** 63-9.
DOING, H. (1972). Proposals for an objectivation of phytosociological methods. *In* Grundfragen und Methoden in der Pflanzensoziologie. Ed. R. Tüxen, The Hague; Dr. W. Junk, N.V., pp. 59-74.

DRISCOLL, R. S. (1971). Color aerial photography; a new view for range management. *U.S. Dep. Agric. For. Serv. Res. Pap. RM* -67.

DRISCOLL, R. S. & FRANCIS, R. E. (1970). Multistage, multiseasonal and multiband imagery to identify and qualify non-forest vegetation resources. *Annu. Progr. Rep. Earth Resour. Surv. Program,* OSSA/NASA, Rocky Mt. For. Range Exp. Sta.

DRISCOLL, R. S., REPPERT, J. N., HELLER, R. C. & CARNEGGIE, D. M. (1970). Identification and measurement of herbland and shrubland vegetation from large scale aerial colour photographs. *Proc.XIth Int. Grassl. Congr.* pp. 95-8.

DYFFEY, E., MORRIS, M. G., SHEAIL, J., WARD, L. K., WELLS, D. A. & WELLS, T. C. E. (Eds) (1974). Grassland ecology and wildlife management. London; Chapman & Hall Ltd.

EVANS, R. A. & LOVE, R. M. (1957). The step-point method of sampling—a practical tool in range research. *J. Range Mgmt.* **10,** 208-12.

FISSER, H. G. & VAN DYNE, G. M. (1966). Influence of number and spacing of points on accuracy and precision of basal cover estimates. *J. Range Mgmt.* **19,** 205-11.

FOSBERG, F. R. (1967). A classification of vegetation for general purposes. *In* Guide to the check sheet for IBP areas. Ed. G. F. Peterken. IBF Handbook No. 4. Oxford; Blackwell Scientific Publications, pp. 73-120.

FRANCIS, R. E., DRISCOLL, R. S. & REPPERT, J. N. (1972). Loop-frequency as related to plant cover, herbage production, and plant density. *U.S. Dept. Agric. For. Serv. Res. Pap. RM*-94.

FREE, J. C., HANSEN, R. M. & SIMS, P. L. (1970). Estimating dryweights of foodplants in faeces of herbivores. *J. Range Mgmt.* **23,** 300-2.

GALT, H. D., OGDEN, P. R., EHRENREICH, J. H., THEURER, B. & MARTIN, S. C. (1968). Estimating botanical composition of forage samples from fistulated steers by a microscope point method. *J. Range Mgmt.* **21,** 397-401.

GALT, H. D., THEURER, B., EHRENREICH, J. H., HALE, W. H. & MARTIN, S. C. (1969). Botanical composition of diet of steers grazing a desert grassland range. *J. Range Mgmt.* **22,** 14-9.

GATES, D. M. (1969). Infrared measurement of plant and animal surface temperature and their interpretation. *In* Remote sensing in ecology. Ed. P. L. Johnson. Athens: University of Georgia Press, pp. 95-107.

GATES, D. M. (1970). Physical and physiological properties of plants. *In* Remote sensing. Washington: National Academy of Science, pp. 224-52.

GIRARD-GANNEAU, C. M. & GIRARD, M. C. (1974). Photographs from balloons: their use in agronomy and management of environment. *In* Proceedings of the IXth international symposium on remote sensing of environment. Vol. II. Michigan : Environmental Research Institute, pp. 1467-74.

GOOD, R. E. & GOOD, N. F. (1971). Vegetation of a Minnesota prairie and a comparison of methods. *Am. Midl. Nat.* **85,** 228-31.

GOODALL, D. W. (1952). Some considerations in the use of point quadrats for the analysis of vegetation. *Aust. J. Sci. B.* **5,** 1-41.

GOODWIN, W. F. & WALKER, J. (1972). Photographic recording of vegetation in regenerating woodland. *Aust. CSIRO: Woodl. Ecology Unit. Tech. Commun. No.* 1.

GRABAU, W. E. & RUSHING, W. N. (1968). A computer-compatible system for quantitatively describing the physiognomy of vegetation assemblages. *In* Land evaluation. Ed. G. A. Stewart. Melbourne: MacMillan of Australia, pp. 263-75.

GREIG-SMITH, P. (1964). Quantitative plant ecology. London: Butterworths Scientific Publications, 2nd Edn.

HANSEN, R. M. & UECKERT, D. N. (1970). Dietary similarity of some primary consumers. *Ecology* **51,** 640-8.

HARGREAVES, J. N. G. & KERR, J. D. (1978). Botanal—a comprehensive sampling and computing procedure for estimating pasture yield and composition. II. Computational package. C.S.I.R.O. Aust., Div. Trop. Crops and Past., Trop. Agron. Tech. Mem. No. 9.

HART, M. L. 't & DE VRIES, D. M. (1949). Grassland and grassland husbandry in the Netherlands. *Rep. Vth Int. Grassl. Congr.,* (Appendix) pp. 1-24.

HAYDOCK, K. P. & SHAW, N. H. (1975). The comparative yield method for estimating dry matter yield of pasture. *Aust. J. Exp. Agric. Anim. Husb.* **15,** 663-70.

HEADY, H. F. & TORELL, D. T. (1959). Forage preference exhibited by sheep with esophageal fistulas. *J. Range Mgmt.* **12,** 28-34.

HEADY, H. F. & VAN DYNE, G. M. (1965). Prediction of weight composition from point samples on clipped herbage. *J. Range Mgmt.* **18,** 144-8.

HERCUS, B. H. (1960). Plant cuticle as an aid to determining the diet of grazing animals. *Proc. VIIIth Int. Grassl. Congr.,* pp. 443-7.

HESLEHURST, M. R. (1971). The point quadrat method of vegetation analysis: a review. *Univ. Reading, Dep. Agric., Study No.* 10.

HESLEHURST, M. R. & WILSON, G. L. (1971). Studies on the productivity of tropical pasture plants. III. Stand structure, light penetration, and photosynthesis in field swards of *Setaria* and Greenleaf *Desmodium. Aust. J. Agric. Res.* **22,** 865-78.

HOFFER, R. M. & JOHANNSEN, C. J. (1969). Ecological potentials in spectral signature analysis. *In* Remote sensing in ecology. Ed. P. L. Johnson, Athens: University of Georgia Press, pp. 1-16.

HOLTER, M. R. (1970). Imaging with nonphotographic sensors. *In* Remote sensing. Washington: National Academy of Science. pp. 73-163.

HUTCHINGS, S. S. & HOLMGREN, R. C. (1959). Interpretation of loop-frequency data as a measure of plant cover. *Ecology* **40,** 668-77.

HUTCHINSON, K. J. (1967). A coring technique for the measurement of pasture of low availability to sheep. *J. Br. Grassl. Soc.* **22,** 131-4.

HUTCHINSON, K. J., MCLEAN, R. W. & HAMILTON, B. A. (1972). The visual estimation of pasture availability using standard pasture cores. *J. Br. Grassl. Soc.* **27,** 29-34.

HYDER, D. N., BEMENT, R. E., REMMENGA, E. E. & TERWILLIGER, C. (1965). Frequency sampling of blue grama range. *J. Range Mgmt.* **18,** 90-3.

HYDER, D. N., BEMENT, R. E., REMMENGA, E. E. & TERWILLIGER, C. (1966). Vegetation-soils and vegetation-grazing relations from frequency data. *J. Range Mgmt.* **19,** 11-7.

HYDER, D. N., CONRAD, C. E., TUELLER, P. T., CALVIN, L. D., POULTON, C. E. & SNEVA, F. A. (1963). Frequency sampling in sagebrush-bunchgrass vegetation. *Ecology* **44,** 740-6.

IBRAHIM, K. M. (1971). Ocular point quadrat method. *J. Range Mgmt.* **24,** 312.

JOHNSON, P. L. (Ed.) (1969). Remote sensing in ecology. Athens: University of Georgia Press.

JOHNSTON, A. (1957). A comparison of the line interception, vertical point quadrat, and loop methods as used in measuring basal area of grassland vegetation. *Can. J. Plant Sci.* **37,** 34-42.

JONES, M. B. & EVANS, R. A. (1959). Modification of the step-point method for evaluating species yield changes in fertilizer trials on annual grasslands. *Agron. J.* **51,** 467-70.

JONES, R. M. (1974). Improvement of the dry-weight rank technique for measuring botanical composition. *Aust. CSIRO Div. Trop. Past. Ann. Rep.* 1973-74, p. 54.

KELLY, B. W. (1970). Sampling and statistical problems. *In* Remote sensing. Washington: National Academy of Science, pp. 324-53.

KEMP, C. D. & KEMP, A. W. (1956). The analysis of point quadrat data. *Aust. J. Bot.* **4,** 168-74.

KERSHAW, K. A. (1964). Quantitative and dynamic ecology. London: Edward Arnold (Publ.) Ltd.

KÛCHLER, A. W. (1967). Vegetation mapping. New York: The Ronald Press Co.

LAMBERT, J. M. & DALE, M. B. (1964). The use of statistics in phytosociology. *Adv. Ecol. Res.* **2,** 59-99.

LEIGH, J. H. & MULHAM, W. E. (1966a). Selection of diet by sheep grazing semi-arid pastures on the Riverine Plain. 1. A bladder saltbush (*Atriflex vesicaria*)-cotton bush (*Kochia aphylla*) community. *Aust. J. Exp. Agric. Anim. Husb.* **6,** 460-7.

LEIGH, J. H. & MULHAM, W. E. (1966b). Selection of diet by sheep grazing semi-arid pastures on the Riverine Plain. 2. A cotton bush (*Kochia aphylla*)-grassland (*Stipa variabilis-Danthonia caespitosa*) community. *Aust. J. Exp. Agric. Anim. Husb.* **6,** 468-74.

LEVY, E. B. & MADDEN, E. A. (1933). The point method of pasture analysis. *N.Z. J. Agric.* **46,** 267-79.

LONG, G. A., POISSONET, P. S., POISSONET, J. A., DAGET, P. M. & GODRON, M. P. (1972). Improved needle point frames for exact line transects. *J. Range Mgmt.* **25,** 228-9.

LYON, L. J. (1968). An evaluation of density sampling methods in a shrub community. *J. Range Mgmt.* **21,** 16-20.

MCCOWN, R. L., TOLSON, D. J. & CLAY, H. J. (1973). Low cost aerial photography for agricultural research. *J. Aust. Inst. Agric. Sci.* **39,** 227-32.

MANNETJE, L. 't & HAYDOCK, K. P. (1963). The dry-weight-rank method for the botanical analysis of pasture. *J. Br. Grassl. Soc.* **18,** 268-75.

MARTIN, D. J. (1964). Analysis of sheep diet utilizing plant epidermal fragments in faeces samples. *In* Grazing in terrestrial and marine environments. Ed. D. J. Crisp. Oxford: Blackwell Scientific Publications, pp. 173-88.

MITCHELL, K. J. & GLENDAY, A. C. (1958). The tiller population of pastures. *N.Z. J. Agric. Res.* **1,** 305-18.

MOORE, R. M. (1970). Australian grasslands. *In* Australian grasslands. Ed. R. M. Moore, Canberra: Australian National University Press, pp. 88-100.

MORRIS, M. J. (1967). An abstract bibliography of statistical methods in grassland research. *U.S. Dep. Agric. For. Serv. Misc. Publ. No.* 1030.

MORRIS, M. J. (1973). Estimating understory plant cover with rated microplots. *U.S. Dep. Agric. For. Serv. Res. Pap. RM*-104.

MOUNTIER, N. S. & RADCLIFFE, J. E. (1964). Problems in measuring pasture composition in the field. Part 3: an evaluation of point analysis, dry weight analysis, and tiller analysis *N.Z. J. Bot.* **2,** 131-42.

NASA (1970). Ecological surveys from space. Washington; NASA, Spec. Publ. 230.

NATIONAL RESEARCH COUNCIL, (1970). Remote sensing. Washington: National Academy of Science.

NUMATA, M. (1966). Some remarks on the method of measuring vegetation. *Mar. Lab. Chiba Univ. Bull. No.* 8. 71-7.

OLUSUYI, S. A. & RAGUSE, C. A. (1968). Inclined point quadrat estimation of species contributions to pasture dry matter production. *Agron. J.* **60,** 441-2.

OOSTING, H. J. (1956). The study of plant communities. San Francisco; W. H. Freeman & Co. 2nd Ed.

OWENSBY, C. E. (1973). Modified step-point system for botanical composition and basal cover estimates. *J. Range Mgmt.* **26,** 302-3.

PARKER, K. W. & HARRIS, R. W. (1959). The 3-step method for measuring conditions and trend of forest ranges: a resumé of its history, development, and use. *In* Techniques and methods of measuring understory vegetation. *U.S. For. Serv., South, For. Exp. Stn. & Southeast. For. Exp. Stn.*, pp. 55-69.

PENFOUND, W. T. (1963). A modification of the point-centered quarter method for grassland analysis. *Ecology* **44,** 175-6.

PHILIP, J. R. (1966). The use of point quadrats, with special reference to stem-like organs. *Aust. J. Bot.* **14,** 105-25.

PHILLIPS, E. A. (1959). Methods of vegetation study. USA: Holt-Dryden.

PIELOU, E. C. (1959). The use of point-to-plant distances in the study of the pattern of plant populations. *J. Ecol.* **47,** 607-13.

PIERCE, W. R. & EDDLEMAN, L. E. (1970). A field stereophotographic technique for range vegetation analysis. *J. Range Mgmt.* **23,** 218-20.

POISSONET, P. S. & POISSONET, J. A. (1969). Etude comparée de diverses méthodes d'analyse de la végétation des formations herbacées denses et permanentes. *Doc. Cent. Nat. Rech. Sci. No.* 50.

POISSONET, P. S., POISSONET, J. A., GODRON, M. P. & LONG, G. A. (1973). A comparison of sampling methods in dense herbaceous pasture. *J. Range Mgmt.* **26,** 65-7.

POULTON, C. E. & TISDALE, E. W. (1961). A quantitative method for the description and classification of range vegetation. *J. Range Mgmt.* **14,** 13-21.

RADCLIFFE, J. E. & MOUNTIER, N. S. (1964a). Problems in measuring pasture composition in the field. 1. Discussion of general problems and some considerations of the point method. *N.Z. J. Bot.* **2,** 90-7.

RADCLIFFE, J. E. & MOUNTIER, N. S. (1964b). Problems in measuring pasture composition in the field. 2. The effect of vegetation height using the point method. *N.Z. J. Bot.* **2,** 98-105.

RATTRAY, J. M. (1960). The grass cover of Africa. *FAO Agric. Studies No.* 49. Rome: FAO.

REPPERT, J. N. & FRANCIS, R. E. (1973). Interpretation of trend in range condition from 3-step data. *U.S. Dep. Agric. For. Serv. Res. Pap. RM*-103.

RICE, E. L. (1967). A statistical method for determining quadrat size and adequacy of sampling. *Ecology* **48,** 1047-9.

RISSER, P. G. & ZEDLER, P. H. (1968). An evaluation of the grassland quarter method. *Ecology* **49,** 1006-9.

ROBERTS, B. R. (1972). Ecological studies on pasture condition in semi-arid Queensland. Brisbane: Queensland Department of Primary Industries, Mimeo.

ROSENFELD, A. (1968). Automated picture interpretation. *In* Land evaluation. Ed. G. A. Stewart, Melbourne: MacMillan of Australia, pp. 187-99.

SCHUBERT, J. S. & MACLEOD, N. H. (1974). Vegetation analysis with erts digital data: A new approach. *In* Proceedings of the IXth International Symposium on Remote Sensing of the Environment, Vol. II. Michigan: Environmental Research Institute, pp. 1193-211.

SHIMWELL, D. W. (1971). The description and classification of vegetation. London: Sidgwick & Jackson.

SHOCKEY, W. R. & DECOURSEY, D. S. (1969). Point sampling of land use in the Washita Basin, Oklahoma. *U.S. Dep. Agric., Agric. Res. Serv. Rep.* 41-149.

SLATER, J. & JONES, R. J. (1971). Estimation of the diets selected by grazing animals from microscopic analysis of the faeces—a warning. *J. Aust. Inst. Agric. Sci.* **37,** 238-40.

SPARKS, D. R. & MALECHEK, J. C. (1968). Estimating percentage dry weight in diets using a microscopic technique. *J. Range Mgmt.* **21,** 264-5.

SPECHT, R. L. (1970). Vegetation. *In* The Australian environment. Ed. G. W. Leeper, Melbourne: CSIRO and Melbourne University Press, 4th Edn., pp. 44-67.

SPEDDING, C. R. W. & LARGE, R. V. (1957). A point-quadrat method for the description of pasture in terms of height and density. *J. Br. Grassl. Soc.* **12,** 229-34.

STEWART, D. R. M. (1967). Analysis of plant epidermis in faeces: a technique for studying the food preferences of grazing herbivores. *J. Appl. Ecol.* **4,** 83-111.

STEWART, D. R. M. & STEWART, J. (1971). Comparative food preferences of five East African ungulates at different seasons. *In* The scientific management of animal and plant communities for conservation. Eds. E. Duffey & A. S. Watt, Oxford: Blackwell Scientific Publications, pp. 351-66.

STEWART, G. A. (Ed.) (1968). Land evaluation, Melbourne: MacMillan of Australia.

STODDART, L. A. & SMITH, A. D. (1955). Range management. New York: McGraw-Hill Book Co. 2nd Edn.

STORR, G. M. (1961). Microscopic analysis of faeces, a technique for ascertaining the diet of herbivorous mammals. *Aust. J. Biol. Sci.* **14,** 157-64.

STORR, G. M. (1968). Diet of kangaroos (*Megaleia rufa* and *Macropus robustus*) and merino sheep near Port Hedland, Western Australia. *J.R. Soc. W. Aust.* **51,** 25-32.

SWITZER, P. (1971). Mapping a geographically correlated environment. *In* Statistical ecology, Vol. 1: Spatial patterns and statistical distributions. Eds. G. P. Patil, E. C. Pielou & W. E. Waters. University Park: Pennsylvania State University Press, pp. 235-69.

TIDMARSH, C. E. M. & HAVENGA, C. M. (1955). The wheel-point method of survey and measurement of semi-open grasslands and Karoo vegetation in South Africa. *Mem. Bot. Surv. S. Afr.* No. 29.

TOTHILL, J. C. & PETERSON, M. L. (1962). Botanical analysis and sampling: Tame pastures. *In* Pasture and range research techniques. Ed. Joint Committee. Comstock Publishing Associates: Ithaca, pp. 109-34.

TOTHILL, J. C., HARGREAVES, J. N. G. & JONES, R. M. (1978). Botanal—a comprehensive sampling and computing procedure for estimating pasture yield and composition. I. Field sampling, C.S.I.R.O., Aust., Div. Trop. Crops & Past., Trop. Agron. Tech. Mem. No. 8.

TREWARTHA, G. T. (1954). An introduction to climate. New York: McGraw-Hill Book Co. Inc.

TUELLER, P. T., LORAIN, G., KIPPING, K. & WILKIE, C. (1972). Methods for measuring vegetation changes on Nevada rangelands. Univ. Nev., Max C. Fleischmann, Coll. Agric., T16.

WALKER, B. H. (1970). An evaluation of eight methods of botanical analysis on grasslands in Rhodesia. *J. Appl. Ecol.* **7,** 403-16.

WARRBN-WILSON, J. (1963). Errors resulting from thickness of point quadrats. *Aust. J. Bot.* **11,** 178–88.

WARREN-WILSON, J. (1965). Stand structure and light penetration. I. Analysis by point quadrats. *J. Appl. Ecol.* 2, 383–90.

WEAVER, J. E. & CLEMENTS, F. (1938). Plant ecology. New York: McGraw-Hill Book Co. 2nd Edn.

WEBB, L. J. (1959). A physiognomic classification of Australian rain forests. *J. Ecol.* **47,** 551-70.

WEBB, J. L., TRACEY, J. G., WILLIAMS, W. T. & LANCE, G. N. (1970). Studies in the numerical analysis of complex rain-forest communities. V. A comparison of the properties of floristic and physiognomic-structural data. *J. Ecol.* **58,** 203-32.

WELLS, K. F. (1971). Measuring vegetation changes on fixed quadrats by vertical ground stereophotography. *J. Range Mgmt.* **24,** 233-6.

WIESLANDER, A. E. & STORIE, R. E. (1952). The vegetation-soil survey in California and its use in the management of wild lands for yield of timber, forage and water. *J. For.* **50,** 521-6.

WILLIAMS, O. B. (1969). An improved technique for identification of plant fragments in herbivore faeces. *J. Range Mgmt.* **22,** 51-2.

WILLIAMS, W. T. (1971). Strategy and tactics in the acquisition of ecological data. *Proc. Ecol. Soc. Aust.* **6,** 57-62.

WILLIAMS, W. T. (Ed.) (1976). Pattern analysis in agricultural research. Melbourne: CSIRO.

WILLIAMS, W. T. & WEBB, L. J. (1969). New methods of numerical analysis of rainforest data. *Malay. For.* **32,** 368-74.

WIMBUSH, D. J., BARROW, M. D. & COSTIN, A. B. (1967). Color stereophotography for the measurement of vegetation. *Ecology* **48,** 150-2.

WINKWORTH, R. E. & GOODALL, D. W. (1962). A crosswire sighting tube for point quadrat analysis. *Ecology* **43,** 342-3.

Chapter 4

MEASURING QUANTITY OF GRASSLAND VEGETATION

L. 't Mannetje

I. INTRODUCTION

This chapter is concerned with methods of measuring the amount of above-ground material present at any one time in plant communities that provide feed for large herbivores. It is the base measurement for nearly all assessments of vegetation. Growth, production, utilization and deterioration are changes in quantity over time. Quantity by quality parameters give a wide range of variables used in grassland research and practice. The base measurement and derived values are mainly used for the following purposes :

(i) to determine the amount of animal feed available;

(ii) to assess the effects of such management practices as the use of fertilizers, herbicides, mixtures of species, choice of grazing methods or stocking rates;

(iii) to calculate growth, utilization by grazing animals, or deterioration.

The same techniques of estimating the amount of vegetation present may be used for each of these purposes, but procedures for sampling and computation will differ, depending on the objectives of the measurements. The value of quantitative measurements for assessing grasslands will depend mainly on the form of utilization. For example, where the vegetation is used for green feeding, hay, silage, or oven-dried products, the quantity coupled with quality in terms of energy, protein and mineral content will determine the value of the vegetation, although losses occurring during conservation, storage and feeding must also be taken into account. However, where the vegetation is utilized by grazing animals, the amount of feed present at any one time is less directly related to its value. In this case it will depend also on method and intensity of grazing and intake by the animals. In addition, animals on pasture are selective in their grazing of plant species and plant parts. This presents a much more complex situation. However, irrespective of whether the method of utilization is by cutting or grazing, the ultimate agronomic value of grassland is determined by animal performance.

The question also arises as to what should be measured. This could be the total vegetation above ground level, or that above a certain height, or certain plant species or plant parts. Vegetation type and method of utilization will determine which is the most appropriate. In the case of open woodlands, for example (Plate 1), only the herbaceous ground flora and perhaps edible shrub components would be important from an animal production point of view. For ecological studies, on the other hand, total vegetative biomass may be more relevant. It is important that these problems are considered at the planning stage of a research programme, in order to prevent the waste of precious resources in obtaining the wrong kind of measurement.

It is impossible and indeed unnecessary to review all the methods that have been proposed over the years. Instead, a selection has been made with the intention of showing the main principles involved. Other reviews on this subject are : Brown (1954), Grassland Research Institute, Hurley (1961), American Society of Agronomy *et al.* (1962), American Society of Range Management and the Agricultural Board (1962), and Milner and Hughes (1968). Morris (1967) published an annotated bibliography on methods of grassland research.

II. SAMPLING

Statistical and botanical aspects of sampling are considered in Chapters 2 and 3. In this section agronomic aspects of sampling procedure and size and number of sampling units will be discussed.

1. Choice of Method

The sampling procedure and intensity depend, first of all, on the scale of the investigation. A different method would be used for small plots for the assessment of pasture species, fertilizer rates, etc., compared to farm-scale paddocks, or areas commonly measured in square kilometers. The method chosen will also depend on the desired accuracy, the purpose for which the data are required and the available resources.

The intensity of sampling is usually inversely related to the size of plots or the area to be sampled, partly because of the time and effort required to complete the work. It is important that sampling for the purpose of comparing treatments or geographical areas is carried out in a reasonably short time to avoid changes being measured which are attributed to treatments or areas. A cost-benefit analysis should be made before embarking on any sampling procedure, particularly where substantial time, facilities or money are involved. The first question to answer is whether the proposed sampling procedure will provide the data required and whether these data can be statistically analysed.

Another question is that of whether the one sampling operation can be used for more than one objective. In addition to quantitative data, the investigation may require some form of quality assessment, such as botanical composition, cover or chemical composition. In this case, there is a choice of whether to carry out all measurements on the same sampling units or to use different sampling units for each kind of measurement. For example, in a grazing experiment one may select a number of sampling units, each one of which is cut for total dry matter, a sub-sample is hand-sorted for botanical composition, and a sub-sample is analysed for chemical composition. Alternatively, total dry matter may be determined using a forage harvester, the botanical composition may be estimated using an indirect method, and samples for chemical analysis may be obtained by plucking.

The sampling procedure selected also depends on whether the data obtained are to be used for a simple descriptive determination of herbage quantity, or for changes in quantity over time, such as growth, utilization, decomposition, or damage. In these cases variability assumes great importance and if comparable precision is to be achieved, more sampling units may be required than in the case of a single estimate (Boyd, 1949). Variability may be reduced by selecting similar paired sampling units, one of which is measured at the beginning and one at the end of the period of study. Paired areas are commonly used, but the author has also used paired plants to reduce the difficulty of selecting equal sampling units. Studies in grazed swards may require the use of exclosures in the form of cages or movable fences. However, the investigator needs to consider the effects of such structures on the vegetation, as will be discussed later.

Many methods are available and the investigator should be aware of their existence, their applicability and their limitations if he is to make the best choice for the particular type of grassland and the problems under study. Factors affecting the choice of method are related to the uniformity, density, height and species composition of grassland, the size and shape of experimental areas, the precision required and facilities and labour available. It is obvious that methods suitable for dense, well fertilized grasslands in humid regions will not be suitable for shrubby open grasslands in arid regions.

Facilities play an important role too; when there is no motorised harvesting equipment available and large areas are to be sampled, one of the indirect methods will be more appropriate than a method relying entirely on cutting samples. In the absence of drying facilities, air-dry weights are to be preferred over fresh-dry weights, because the latter depend on weather conditions, time of day and plant species.

The required precision of estimation depends on the objective of the research. For example, yield may be one factor used to describe various forms of vegetation in which case only crude estimates of yield will be necessary. On the other hand, a comparison of cultivars of a given species will require a very precise method and the use of a sampling design aimed at reducing errors to a minimum may be called for, because differences among cultivars are likely to be small. An understanding of the type and magnitude of errors and bias is essential to make the best choice of method within the limitations of facilities and labour (see Chapter 2).

2. Choice of sampling unit

The size, shape and number of sampling units depend on the degree of accuracy desired, the type of vegetation, method of harvesting, available facilities, labour and so on. Apart from statistical aspects discussed in Chapter 2 there are biological, practical and economic considerations to be taken into account and the final choice of sampling units is usually a compromise between accuracy, practicability and cost.

Sample shapes in use are square, rectangular and circular. Both the frame used to delineate the area and the area itself are commonly called a *quadrat*, irrespective of the actual shape. However, when the area is many times longer than it is wide, and when cutting is done by power-driven equipment, the sampling unit is often referred to as a *strip*. The material used to construct frames may be wood or metal, but light-weight aluminium is preferred for ease of construction and for its non-rusting quality. Sample units may also be delineated by flexible material such as string or plastic lines. When cutting with a power-driven blade of a known width it may be necessary only to measure the length of the strip. In tall or very dense vegetation, one side of a four-sided quadrat frame is often left open for ease of placement. Circular quadrats may be difficult to place into position in any but short grassland. To overcome this problem Kennedy (1972) developed the ' sickledrat ', which consists of a stake which is pushed into the ground and which forms the centre of the quadrat. An arm with a curved, pointed cross-piece at the end rotates around the stake. Round sample units in the form of turf corers have also been used for very short vegetation. De Vries (1937, 1940) described a tapered corer 10 cm high with a diameter of 56 mm at the bottom, and of 66 mm at the top, fitted with a cross handle for use in dense mixed *Lolium perenne* grassland. Hutchinson (1967) used the pasture sampler described by Mitchell and Glenday (1958) and studied the effects of core size and shape on *Phalaris tuberosa* and *Trifolium repens* pastures.

The choice of shape and size of quadrats is also influenced by the method of harvesting. Self-propelled or tractor-driven implements are best used on strips. Because of inaccuracies in starting and stopping, errors caused by this will be minimised using a strip with a large length : width ratio. Small quadrats of any shape and circular quadrats do not lend themselves to cutting by large motorised equipment and are best cut by hand-held tools, which may be power operated.

A large proportion of sampling error arises from edge effects. They may be due to difficulties of boundary definition, disturbance of the vegetation by the frame by either pushing material in or out of the area. It has to be decided beforehand whether to include only herbage that originates from inside the quadrat, or to include material actually present within the vertical planes of the quadrat sides, irrespective of its point of origin. With stoloniferous or creeping material it is advisable to choose the latter method as it is more accurate and quicker; for short, dense grassland this is the only possible way as it is not feasible to separate plant units. In tall vegetation this problem also applies to the vertical boundaries. Plants do not usually grow upwards in strictly vertical planes and material growing both within and outside the quadrat will cross the boundaries. With twining species such as many tropical legumes, care should be taken that material is not pulled inside the quadrat, thus giving an overestimate of yield for that species. To reduce edge effects, the quadrat perimeter should be as small as possible in relation to the area of the sample unit. The perimeter to area ratio is smallest for a circle and for this reason Van Dyne, Vogel and Fisser (1963) and others have recommended the use of circular quadrats. However, owing to difficulties of placement, except in the case of the sickledrat mentioned earlier, it is more practical to use an open-ended square or rectangular quadrat, the latter also being favoured on statistical grounds (Chapter 2).

Because quadrat size and number are the main factors influencing accuracy, these have been the subject of many investigations. However, total area cut also determines the effort involved in sampling. To help decide the best number and size of quadrats, the cost : benefit ratio must be considered. A method of determining the optimum number of sampling units that is non-statistical and easy to use is given by Greig-Smith (1964). The mean of the variable being estimated (yield) is plotted against the number of quadrats. The fluctuation in mean values will decline as the number of quadrats from which the mean is calculated is increased. From such a plot and the time and effort required to harvest the quadrats, a decision can be made as to the most acceptable number.

Random quadrat location is described in detail in Chapter 2. A form of bias often overlooked is that associated with operators, which may result from quadrat placement and cutting technique, differences in observation, or fatigue. To avoid this, each operator should deal with an equal number of sampling units in each treatment, or in replicated experiments each operator could do all the work in one replicate. Differences between operators would thus be eliminated or confounded with block differences.

III. METHODS OF MEASURING QUANTITY OF VEGETATION

The methods available can be grouped broadly into *destructive* and *non-destructive* ones, but all require some form of cutting. The difference between the two groups of methods is that in the first group, the amount of vegetation of an area is estimated by cutting techniques only, whereas methods used in the second group usually involve the measurement of one or more variables that can be related to quantity by the destructive harvesting of only a small number of sampling units. The methods of this group are termed *non-destructive*, because the indirect measurements can be made on observational plots, leaving the vegetation intact for further experimental treatments, and the direct measurements to be made on the vegetation outside, provided it is the same as that in the observational plots.

The various methods can also be regarded as a progression of reduced physical effort. Starting with the complete harvesting of a plot, or of a field as is done for fodder conservation, the work effort can be reduced by harvesting a number of sampling units only. For various reasons this may not be desirable, either because the physical effort is still too great, or the vegetation has to be measured without interference. In this case a double-sampling technique is introduced. There is a vast number of methods available for the different categories. No one method *per se* is more or less accurate than another. Any given method can be judged only in relation to the type of grassland, its utilization, and particularly the statistical design of the sampling procedure. (See Chapter 2).

1. Destructive techniques

(a) Cutting equipment

The simplest harvesting devices are hand operated tools such as scissors, shears, secateurs, knives, sickles and scythes. Although these require a high labour input, they are useful with small quadrats or with irregular or round ones. Scissors, secateurs and shears have the distinct advantage over all other cutting devices that cut material can be separated into its components in the field, where individual plants are easily distinguishable. Another advantage with these is that height of cutting can be more accurately controlled. However, mechanization of yield sampling has resulted in the adaptation of power-driven equipment and the development of special machines. These range from hand-held sheep shearing devices and lawn mowers, to full-size tractor-driven or self-propelled mowers, either with reciprocating knives, rotary blades or flails. Types of power-driven, hand-held tools are the sheep-shearing hand pieces (Alder & Richards, 1962; Jones, 1973) and the cordless hedge-trimmer (Matches, 1963).

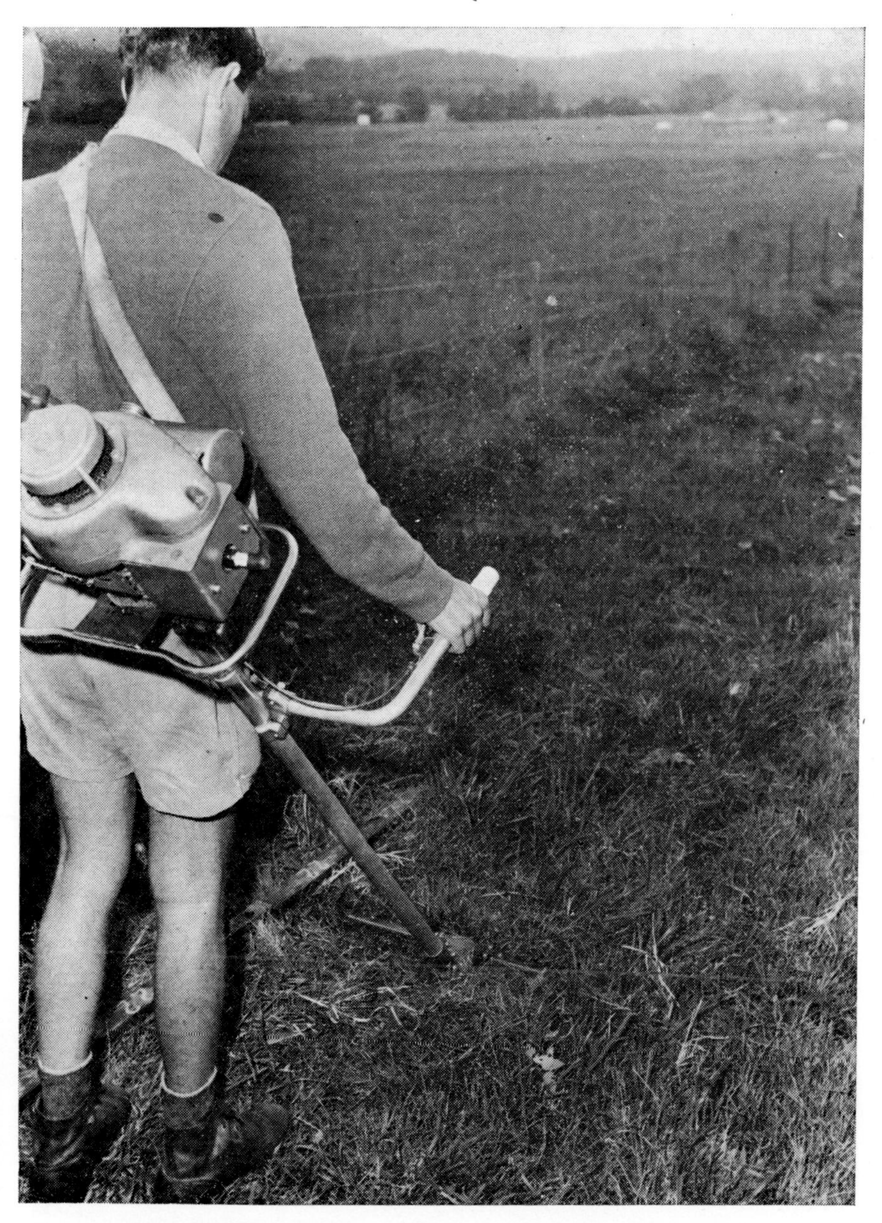

PLATE 6. Petrol-driven portable power-scythe and measuring stick

Van Dyne (1966) incorporated a vacuum cleaner with a sheep-shearing handpiece to collect the material, but although it resulted in better recovery of cut herbage, it also collected some litter and soil (see below). Intermediate-sized machines in use are commercially available, portable power scythes or hedge trimmers (Hedrick & Hitchcock, 1953) which have a reciprocating cutter bar 50 to 75 cm long at the end of a power shaft which is driven by a small petrol engine; they may be equipped with wheels near the blade or a sling for carrying over the shoulder (Plate 6). Some also have detachable height adjusters. A somewhat bigger machine of this nature is a self-propelled auto-scythe with a cutter bar 90 to 100 cm long equipped with a gear box. The disadvantage of all hedge trimmers and power scythes is that they are not equipped with collecting trays. However, these can quite easily be attached behind the cutter bar of auto-scythes, but not hedge trimmers. An example of a collection tray is shown in a photograph by Matches (1966).

Rotary lawn mowers equipped with a collection bin are frequently used for yield sampling (e.g. McGinnies, 1959), but they are unsuitable in tall vegetation. Mowers of three types are in use : (1) walk-behind, push, (2) walk-behind, self-propelled, (3) Sit-on, self-propelled. Some self-propelled mowers do not

PLATE 7. Tractor-mounted forage harvester with collection bin and clock-face scales. Note winch for lowering and raising the bin for weighing

have separate drives for mowing and for propulsion; these are unsuitable for sampling because they cut a swath when driving to the next sampling position.

Forage harvesters in use for sampling vary in size and design from small self-propelled to large tractor-mounted machines (Plate 7). They need to have provision for collecting and weighing material, which nearly always requires modifications to commercially available machines. Numerous innovations and modifications to forage harvesters have been proposed to obtain satisfactory performance for plot harvesting, some of which have been reported by the following : Howell (1956), Fortmann (1956), Kemp and Kalbfleisch (1957), Patterson and Browning (1962), Thompson and Heinrichs (1963), Buker (1967), Swallow (1967), Allen, Casselman and Thomas (1968), Collins, Rhykerd and Noller (1969), and Martin and Torssell (1971). None of these devices is applicable to very short pastures, for which special coring techniques are more suitable, as discussed in the previous section (see pp. 66 and 67).

(b) *Area sampled*

The total area harvested may range from 100 percent on small plots to 0·5 percent or less on very extensive areas. In small-plot work the whole plot may be harvested, but it is advisable to discard a border area to eliminate edge effects. Alternatively, a proportion of the plot may be harvested from the centre or from random positions. Small plots are usually sampled with hand-held tools, but a row of plots may be sampled by a continuous mower strip cutting across all plots.

Cut herbage is sometimes returned after weighing and sub-sampling for dry matter, botanical or chemical analysis. This may be beneficial in small plot experiments where removal of all the material could affect the nutrient cycle, but large masses of cut material may impede regrowth, particularly in dry conditions (Sears, 1951). As the area to be sampled increases, the proportion harvested usually decreases. However, with the use of machinery and a well designed sampling procedure it is possible to harvest within a reasonable time an area of 1 to 2 percent of experiments covering 100 ha or more.

(c) *Height of cutting*

A basic requirement for all equipment used for cutting samples is the ability to control the height of cut. The basal portions of grassland plants are heavy and, in short grassland, may contain a very significant amount of animal feed. Hand held devices allow for cutting to ground level, which is recommended because then the true total above-ground biomass can be estimated. Another important consideration is that cutting to any height above ground level may cause bias between treatments cut by one person, or bias between persons. However, cutting to ground level may damage the vegetation and introduce an unnatural factor influencing regrowth. Recurrent samples should therefore not be taken from the same position over short time intervals. As a rule, power driven equipment, except sheep-shearing hand pieces, cannot cut to ground level, but their cutting height must be independently controllable. Because of its contribution to the total yield, the material below cutting height should be estimated separately and this should be added to the yield estimated by machine harvesting. This can be done by sampling the stubble by means of hand-held tools.

(d) *Weighing and sub-sampling*

Material from small quadrats can be dried and weighed without sub-sampling. However, where the amount of herbage is large, the material should be weighed fresh, thoroughly mixed and sub-sampled for dry matter determination. With long material the mixing will be facilitated by chopping. Sub-samples of about 1 kg are usually weighed in a laboratory and care should be taken that there is no loss of moisture before weighing. The use of plastic bags or tins with a sealed lid is essential. Even so, all samples should be dried as soon as possible after cutting to avoid further dry matter losses and changes in chemical composition (see Chapter 5).

(e) *Sample contamination*

Herbage samples obtained by any mechanical device that incorporates herbage collection are likely to be contaminated with litter, soil and dung. This is particularly so for material cut close to ground level and with machines that incorporate suction, such as forage harvesters. Not only does this give rise to over-estimation of the dry matter present, but will also interfere with chemical analysis of the herbage. The simplest way to prevent errors arising from such contamination is to collect samples for chemical analysis by hand-plucking, independently from the yield sampling. Alternative methods for avoiding or overcoming these complications are discussed in Chapter 5. (See also Thompson & Raven, 1955). Hebblethwaite and Hughes (1959) described a method to correct weight errors in soil-contaminated samples, using ashing.

2. Non-destructive techniques

Harvesting of quadrats or strips requires high inputs of labour and sometimes equipment also. It is therefore a costly operation and costs may be prohibitive when many treatments or large fields are to be sampled. Investigators may be tempted to compromise by taking fewer sampling units, but this may lead to inadequate sampling, resulting in low precision. Destructive methods may also not be applicable in certain circumstances, for example, where yield is to be determined without interference at intervals on the same area, and in small grazed plots where the total amount of herbage harvested would be too large a proportion of available feed. These reasons have led to the development of a large number of non-destructive methods, which may be grouped as follows :

(i) eye-estimation,
(ii) height and density measurements,
(iii) measurements of non-vegetative attributes, such as capacitance, beta-attenuation and spectral composition of reflected light.

Although these methods are not as accurate as cutting on a per sample basis, they frequently offer the possibility of a net increase in precision because of a large increase in sample number. They also obviate the cutting height problems which occur with all cutting methods.

(a) *Eye-estimation*

In its simplest form an observer makes an estimate of the total amount of herbage present or of the annual production potential of a field, without any checks on actual yields. Although there are observers who possess such ability to a high degree, the procedure is of doubtful value in critical research, because it is entirely subjective and lacks repeatability. However, for the purpose of large-scale surveys, such as grassland mapping in the Netherlands (De Boer, 1956), eye-estimation coupled with actual yield sampling on control plots has proved most useful. Ferrari (1953) compared errors from eye-estimation with those from cutting-estimation for yield and found that errors for the eye-estimates were only 1·7 times those of cutting.

The first visual estimation method acceptable in critical research was the weight-estimate of Pechanec and Pickford (1937), whereby observers estimate the yield of individual plants after a long period of training involving repeated checks of subjective estimates against actual weights. Hutchings and Schmautz (1969) introduced the relative-weight estimate on the assumption that there is less room for errors when yield is expressed as a percentage of that of a nearby standard than when estimating by eye in terms of actual weight. The procedure they used was to locate a quadrat of about 30 cm $\times$ 60 cm at random and regard it as the centre of a square. At each corner of this square another quadrat of the same size was placed; quadrat centres were about 1 m apart. Quadrat size and distances between quadrats may of course be varied. The yield on the corner plots is estimated as a percentage of the centre plot, which is harvested, dried and weighed.

The methods of Pechanec and Pickford and of Hutchings and Schmautz may contain personal bias from visual estimates. Wilm, Costello and Klipple (1944) applied the double-sampling technique to the weight-estimate method. As the term implies, double-sampling involves two methods of sampling the same population. One is the accurate determination of yield (or any other attribute) in a few samples (standards) and the other is an eye-estimate of yield (scores) in many samples, including the standards. For each observer a regression is calculated between actual yield and visually estimated yield of the standards. This regression is plotted and the yield of the field is calculated by converting the scores of the visual estimations into actual yields. Tiwari, Jackobs and Carmer (1963) described the statistics of double-sampling (see also Chapter 2). Morley, Bennett and Clarke (1964) applied double-sampling to a large grazing experiment in southern Australia. A similar method was used by Symons and Jones (1971) in South Africa, and by Haydock and Shaw (1975) in Queensland.

Campbell and Arnold (1973) tested the method of Morley, Bennett and Clarke (1964) in a range of green and dry pastures varying widely in botanical composition and yield. They concluded that at least ten standards that include the highest and lowest yield per quadrat encountered should be used, and that observers should be trained. This training may include the use of colour slides to show certain forms of bias. In the method of Haydock and Shaw (1975) the yield of a quadrat is related to that of a set of reference quadrats, whereas Morley, Bennett and Clarke (1964) estimated quadrat yield directly. These methods are time consuming in the tabulations and calculations required and the use of a computer programme will greatly improve their efficiency, particularly if field data are directly written onto computer data sheets, thus obviating the need of transcription.

For very short pastures (less than 20 cm) Hutchinson, McLean and Hamilton (1972) adapted the coring method of Hutchinson (1967) as an eye-estimation technique. Eight standard cores representing the full range of pasture yields in terms of green herbage are selected and assembled in order on a circular tray with a hole in the centre of the same diameter as the standard core. The tray is placed on a number of randomly selected sites and observers score the vegetation on a scale of 0 to 8, 0 representing bare ground and 1 to 8 representing the standards on the tray. On completion of the observations the yield of the standards is determined by dissection of the cores and the scores of the sampling sites are converted to actual yields. A double-sampling technique to account for personal bias is proposed by the same authors; in this procedure the standards on the tray are used only as a visual aid for scoring. A proportion of the random sites is cored and the yield determined for use in a regression technique as described above.

Wells (1971) pointed out that vertical ground stereo photography could be useful in eye-estimation techniques. A number of reference photographs depicting the standards could be consulted with a pocket stereo viewer, which might increase the accuracy of estimation.

(b) *Height and density*

The yield of an area of vegetation is related to the density and height of individual components. Brown (1954) reviewed early methods based on this relationship, but numerous new ones have been proposed since. No single method could be expected to be satisfactory for all vegetation types and a number of alternatives will be briefly discussed here.

On an individual-plant basis Hurd (1959) found that 86 to 94 percent of the variation in herbage weight of *Festuca idahoensis* could be accounted for by combining measurements of maximum leaf height, basal area and number of flower stalks. Leaf height was the best single measure related to dry weight.

Ground cover and sward height have been used on different types of grassland to estimate relative dry matter yield, e.g. Pasto, Allison and Washko (1957), Spedding and Large (1957), Evans and Jones (1958), Bakhuis (1960), Hughes (1962) and Alexander, Sullivan and McCloud (1962). Symons and Jones (1971) discussed various non-destructive methods and pointed out that attributes such as height and density of vegetation are not easily defined and are therefore subject to error and bias. Plant height, for example, is subject to wind and lodging and if the height of vegetation in an area is to be multiplied by the density in that area, difficulties arise as to how height should be measured. Dann (1966) used a ruler to measure height in small quadrats in a *Paspalum dilatatum* and *Trifolium repens* pasture of uniform density. He obtained a correlation coefficient of 0·95 between sward height and dry matter yield, but in a variable stand of *Sorghum* species this value was reduced to 0·71. Some objectivity may be achieved by using a grass-height measuring disc (e.g. Alexander, Sullivan & McCloud, 1962; Jagtenberg, 1970) and a point quadrat frame for estimating cover (Chapter 3). Weighted discs may be used on grassland of homogeneous composition to measure daily increments in yield, as was shown by Phillips and Clarke (1971). Whitney (1974) developed an apparatus for measuring sward height of *Pennisetum clandestinum* and *Digitaria decumbens*, two low growing grasses. It consists of a fresnel lens attached to a PVC pipe which slides over a steel pipe (Plate 8). The device is put in position and the fresnel lens is lowered on to the sward and then lifted until only the tips of most of the leaves are touching the lens. The lens height can be read on a scale on the upper portion of the steel pipe. Correlation coefficients of 0·97 were obtained between height and weight. The linear regression equations were different for the two grasses, but similar for *P. clandestinum* at different sites.

For rows of one species Woodhouse, Peterson and Berenyi (1963) developed a device which measures the resistance to compression in a given length of row. It consists of two wooden compressor bars, one with a pulley, the other with a cord and a weight at the end. The bars slide on aluminium cross bars provided with pointed legs. The device is put across a row, the cords put over the pulley and the final distance between the compressor bars is a measure of the resistance of the vegetation between the bars, which is proportional to the volume. The method has been used successfully in dune stabilisation work and is of general interest for rows of sown grassland.

To convert any of these plant measurements into actual yield, calibration tests must be carried out. It has been pointed out by several authors that these techniques are not always quicker or less tedious than cutting techniques, but

PLATE 8. Fresnel lens apparatus for measuring sward height (Whitney, 1974)

for comparative purposes they can be truly non-destructive. However, they can be used only on monospecific swards of uniform density. Best results will be obtained when the height of the sward is not too variable.

(c) *Capacitance*

Fletcher and Robinson (1956) were the first to apply the capacitance method to measuring herbage yield. The principle is that air has a low and herbage a high dielectric constant. A meter measures the change in capacitance caused by replacing air by herbage under a measuring head. The first meter, consisting of three parallel aluminium condenser plates, was put on to the herbage, and the capacitance was measured directly. Considerable modification to the device and the development of special sampling techniques took place at a rapid rate. Most present day devices are modifications after Campbell, Phillips and O'Reilly (1962), who used probes encased in insulating plastic tubes mounted on metal bars as the measuring head. The frequency of a valve oscillator on top of the measuring head was controlled by the capacitance under it. This frequency was measured with a hand-held radio-frequency oscillator by an operator with earphones a little distance away. By changing a dial a sharply defined nul-point was established which was read on a scale. This reading could be converted to herbage yield after calibration. Bryant *et al.* (1971) obtained disappointing results with a capacitance meter similar to that of Campbell, Phillips and O'Reilly (1962). They established 84 regressions between meter readings and pasture yield over a two-year period, but found that only 44 of these were significant at the $P = 0 \cdot 05$ level.

Jones and Haydock (1970) described further modification which resulted in the development of a commercially available 'pasture meter' in Queensland, Australia (Plates 9, 10). There is a choice of two heights, one for short and one for tall pastures, both measuring to ground level, and the electronic equipment is fully transistorised and attached to the meter for direct reading. There are two scales, one for low and one for high pasture yields. Jones and Haydock developed the following sampling technique : a zero reading is obtained from an area clipped to ground level, and a number of readings at random positions on the vegetation is obtained. The mean reading is calculated and three positions are found with this mean reading. Each of these is cut, dried and weighed, resulting in three independent estimates of the mean yield.

Neal and Neal (1973) reviewed and described the technical development of capacitance meters. Further testing of a capacitance meter was done by Johns (1972), particularly with regard to double-sampling techniques, and he also indicated the influence of season, species and irrigation on meter calibration.

Jones and Haydock (1970) found that water has the predominant influence on meter readings. Desiccation of herbage produced low readings and application of water to herbage increased readings. Herein lies the main problem with the use of the meter for estimating pasture yield. Dry, standing herbage contains little moisture; it therefore has little capacitance and meter readings are liable to be inaccurate. Pastures consisting of different species may pose problems when the components have different moisture contents. In this connection, Currie, Morris and Neal (1973) working in a winter-rainfall area of low precipitation in Arizona, USA, found that readings with an American capacitance meter accounted for only 47 percent of an *Eragrostis curvula* pasture of mixed composition. Removal of readings related to identifiable areas dominated by a shrub, by dead material or by very succulent species improved this figure to 54 percent. However, on green monospecific vegetation such as nitrogen-fertilized grassland, or fodder crops such as oats, the meter may be used to great advantage, provided

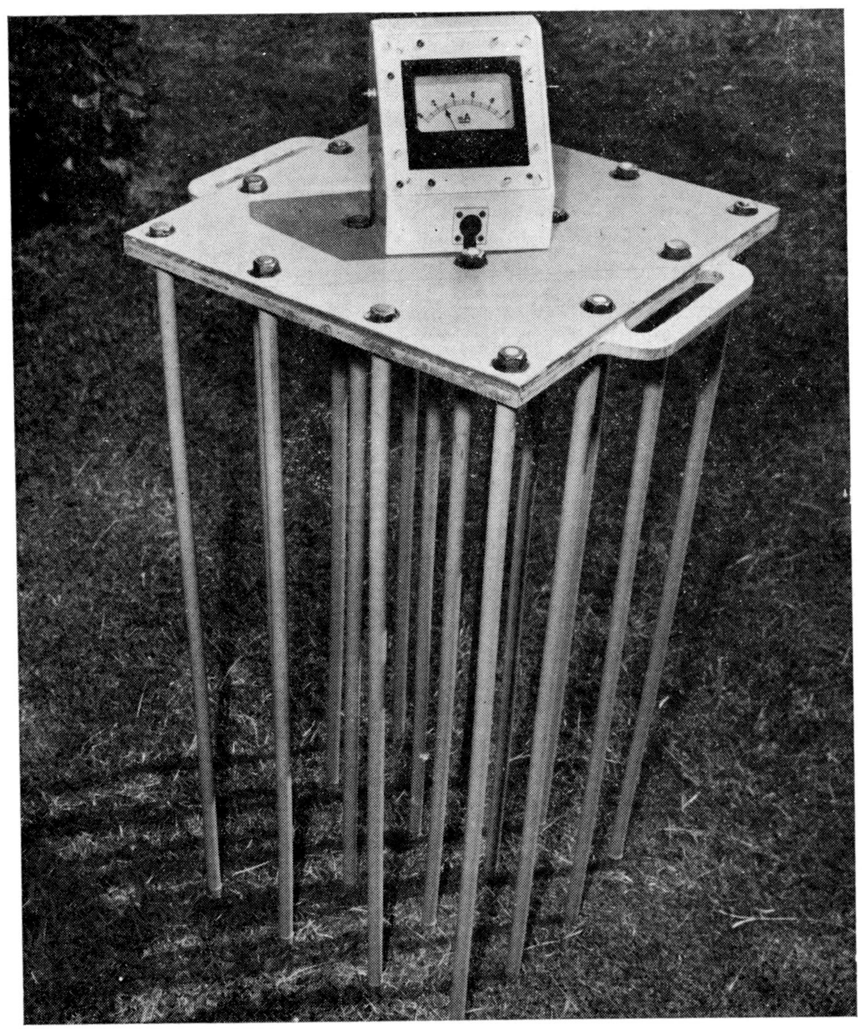

PLATE 9. Electronic pasture meter (Jones & Haydock, 1972)

changes in moisture content such as occur diurnally and between treatments are taken into account.

Carpenter, Wallmo and Morris (1973) used a capacitance meter to estimate the amount of leafy material in shrubby *Artemisia tridentata* vegetation. Wood has little capacitance and it was found that mature woody stems could be ignored. A similar development is being followed by Jones and Sandland (1974) for subtropical grasslands. They found that where a correlation coefficent of 0·65

PLATE 10. Electronic pasture meter mounted on frame with bicycle wheels for one-man operation. When the handlebar is lowered the meter swivels between the wheels

existed between meter reading and total dry matter yield, it was 0·85 between meter reading and green material on a dry matter basis.

(d) *Beta-attenuation*

Teare (1963), Mott, Barnes and Rhykerd (1965) and Teare, Mott and Eaton (1966) applied the principle of thickness-gauging by radio-isotope attenuation to estimate bulk density of standing herbage. By integrating bulk density over height a measure of mass per unit area is obtained which can be related to dry matter yield. A ' foliage-meter ' was developed using radio-active material with a long half-life (strontium-90 or chlorine-36). Published tests of this device have been made with chromatography paper, wooden dowel rods and plant material grown in boxes. The radiation detected from strontium-90 over a distance of 45 cm through soybean and wheat predicted green weight accurately, but the linear regression lines for the two crops had different slopes. There was a strong effect of moisture on attenuation when wooden dowel rods were soaked in water.

The first field application was made by Mitchell (1972) in short-grass prairie. The equipment consisted of an open-ended cardboard box, a Geiger-Mueller tube, portable scaler, radio-active source and a survey meter. The correlation coefficients between beta attenuation and herbage yield in 35 samples were between 0·90 and 0·94. Variability was greatest at lowest yields and the method was not recommended for very short vegetation. Micro-relief of soil surface would also affect the reading, as some herbage might be sheltered in slight depressions. Due to the random emission of beta particles, measurements should be made to record the accumulative activity over a period of about 1 minute. The calibration requirements were not determined.

(e) *Spectral analysis*

Short-wave radiation transmitted and reflected by vegetation differs in spectral composition depending on vegetation type and ratio of dead to green material (Scott, Menalda & Brougham, 1968). Working within the U.S. International Biological Program (IBP) Grassland Biome, Pearson and Miller (1973) developed a hand-held biomass meter (' biometer '). The yield estimate is based on the ratio between two radiance or reflectance measurements at two wavelength bands. The prototype of the biometer consists of a digital radiometer, a pocket calculator and an electronic circuit powered by a 15-V battery pack to connect the two instruments and to solve the biomass estimation equations. The proposed sampling technique is to measure the channel ratio of a few plots of which the yield is determined by clipping. This gives a linear relation between channel ratio and yield which is then entered into the calculator. The biometer can then be used on a large number of sampling units of the same vegetation. In their tests Pearson and Miller obtained a correlation coefficient of 0·98 between reflectance ratio and green yield from 20 plots of short-grass prairie. When the linear regression equation was used predictively on 20 unknown plots of the same field, the correlation coefficient was 0·95. They pointed out that high yielding grassland vegetation types may require non-linear equations due to differences in radiances in various layers of vegetation. Further tests and the construction of a commercially available meter are in progress.

3. Special techniques for shrubs and trees

Grasslands as defined in Chapter 1 may contain woody plants. These vegetation types are found all over the world, for example in the hills of the UK (the ' heather moors '), in rangelands of the USA and Canada, in the semi-arid regions of Australia and in the wet tropics. Shrubs or trees are sometimes planted for animal feed, e.g. *Leucaena leucocephala*.

There are two fundamentally different approaches. Firstly, the determination of total amount of dry matter above ground level may be the objective, in which case all herbaceous and woody plants are harvested completely. Secondly, it may be desired to measure only the material edible to livestock or wildlife. The main difficulty that arises with this approach is to define what is edible. Not only will this differ between plant and animal species, but it will also vary with the prevailing circumstances, which may include drought, resulting in a general shortage of feed. Then there is the problem of sampling units. Should this be an area of ground or a number of plant units of which some defined attribute is measured ?

In this section a number of alternative methods for the determination of yield of shrubs and trees will be mentioned, but their applicability will depend on local conditions. Total biomass of stands of *Leucaena leucocephala* has been measured by cutting to 5 to 10 cm above ground level using a motor scythe (Takahashi & Ripperton, 1949; Anslow, 1957) but the leaves and small twigs are of most interest for animal feeding. Brown (1954) based the productivity of shrubs on " the portion of the current year's twig growth which is edible and available to animals ". An arbitrary, but well defined technique was used by Hutton and Bonner (1960) with *L. leucocephala*. They harvested leaves and twigs to a diameter of 6 mm (1/4 inch). R. J. Jones of the CSIRO Division of Tropical Agronomy, Brisbane, Australia, used a piece of steel with a loop at one end with an internal diameter of 6 mm to indicate the point at which the twig is to be cut. Secateurs are the best hand tools for this technique.

Grant (1971) studied the measurement of primary production of heather (*Calluna vulgaris*) in Scotland. Total green dry matter was used as a measure of edible dry matter. This was determined by estimating yield by destructive sampling of quadrats and hand separation of random samples of plants into flowers, current season's shoots, older green material, dead material and woody stems. To reduce the labour involved, chlorophyll extraction techniques were applied as described in the next section.

Non-destructive methods have also been used to estimate yield of shrubs. Hormay (1943) measured the diameter of the crown of bitter brush in California and multiplied crown area by average estimated twig length to give a relative yield score to use in comparison between shrubs or rows of shrubs. Burrows and Beale (1970) studied *Acacia aneura* (mulga), a very important fodder tree in semi-arid Australia. By measuring the stem circumference at 30 cm above ground level and weighing the leaves after drying, they found a correlation coefficient of 0·98 for the linear relation between these variables expressed as logarithms. Provided that the trees sampled are representative of the size distribution of the whole population, this appears to offer the possibility of using the relationship in a double-sampling technique for leaf yields of single stemmed fodder trees.

In the section on capacitance it was already mentioned that Carpenter, Wallmo and Morris (1973) used a capacitance meter to estimate leaves in a short-shrub vegetation.

4. Composition of yield and percentage green material

Total dry matter yield is of little use in studies of mixed vegetation unless it is further qualified. In Chapter 3 of this book Tothill has outlined methods for determining botanical composition. Botanical composition by weight can either be determined by indirect methods on the standing herbage, or by hand separation of cut material. Indirect methods give the percentage composition, and the contribution to yield by the various components is then calculated using total yield data. Hand separation usually requires sub-sampling because the total amount of cut material is too great to handle. The sub-samples may be taken before or after cutting for yield sampling. Certain types of machines may lacerate the herbage to some extent, making hand separation difficult. In this case, it is better to cut small quadrats by hand. The material so collected may be sorted in the laboratory within a short period or stored in chilled conditions for sorting later. An alternative is to separate during cutting by hand. This can only be done when the components are present in clearly distinguishable units such as in open grassland, or where the components differ greatly in growth form.

It is often desirable to distinguish between green material and dead material. This can be done by hand separation, which is very time consuming, or by estimation based on separation of a few standard or check samples. The estimation can be made in the field, or in cut samples before drying. Some alternative ways are as follows:

(i) The ' constituent differential ' method of Cooper *et al.* (1957), which was developed for the determination of species composition by weight of a two-component herbage mixture where the components differ in the concentration of some constituent. The constituent concentration of a sample of the total herbage and that of each of the components is determined and the composition calculated as follows :

$$X = 100 \frac{C_T - C_1}{C_2 - C_1}$$

where X is the percentage composition by weight of component 2, C_T the concentration of the constituent in the total sample and C_1 and C_2 the concentration of the constituent in components 1 and 2, all as a percentage of the dry matter. Hunter and Grant (1961) used the method to determine the proportions of green and dead material and used either percentage dry matter or percentage crude protein as a constituent.

(ii) The ' pigmentation method ' of Hunter and Grant (1961). Ground, oven-dried green and dead samples are mixed by weight to cover the full range from dead to green at 10-percent intervals. Methanol is added and the extract filtered off. Colour density, as measured by an absorptiometer, setting the 100 percent dead at 100 percent light transmission, is plotted against percentage green to give the calibration graph, which is nearly linear. Samples of the ' unknown ' are treated in the same way and by means of the calibration graph the percentage ' green dry matter ' is determined.

(iii) The ' chlorophyll extraction method ' of Grant (1971) is similar to the pigmentation method except for the preparation of the extract, which is done in subdued light with acetone and a trace of crystalline Na_2CO_3. Optical density of the standard and the unknown samples are read on a spectrophotometer. For both extraction methods new calibration graphs must be prepared for different times of the year and for different stands. The chlorophyll method gave a correlation coefficient of 0·99 between percentage green material by extraction and that by separation.

(iv) The ' dry-weight rank method ' of 't Mannetje and Haydock (1963), developed for botanical analysis of grassland vegetation *in situ* (Chapter 3). Components in a randomly placed quadrat are ranked in order of dry weight into first, second and third. The method can be used for estimating green and dead material by mentally splitting each component into these fractions.

Of the various methods described the dry-weight-rank method is the quickest but is possibly less accurate than the others. Hunter and Grant (1961) compared hand separation with methods (i) and (ii) described above. They found that the constituent differential method using dry matter percentage was acceptable and was the fastest, but pointed out that this can only be used when the vegetation is free of adhering moisture. Use of crude protein as the constituent did not give acceptable results compared with hand separation. The pigmentation method was an adequate substitute for hand separation, provided that the yield samples were reasonably homogeneous in botanical composition. A double-sampling procedure was suggested to correct the pigmentation method, using hand-separation in some samples.

IV. GROWTH AND UTILIZATION

So far in this Chapter only the standing biomass at any one time has been considered. This is the end-result of past growth minus utilization and decay. Growth and utilization are changes in biomass over time. This statement may give the impression that growth and utilization can simply be defined as the difference between yield determinations at the beginning and the end of a period. This may be so over very short time intervals, but it does not take into account any undetected changes during the interval. These changes may be due to the disappearance of material present at the beginning, or the production of new material that disappears again before it is measured at the end, either through senescence, environmental stress or consumption by wildlife and insects. Wiegert and Evans (1964) defined growth (Y) with the exclusion of livestock, wildlife and insects as $Y = \triangle G + X + \triangle D$, where $\triangle G$ and $\triangle D$ are the measured changes in green and dead standing crop during the time interval, and X is the amount of dead material disappearing during the time interval. To determine Y, five variables must be measured: (i) dead material at the start, (ii) dead material at the end, (iii) green material at the start, (iv) green material at the end, and (v) instantaneous daily rate of disappearance of dead material during the interval. Modifications to this method were proposed by Lomnicki, Bandola and Jankowska (1968) and by Wallentinus (1973), but Singh, Lauenroth and Steinhorst (1975) criticised the underlying assumption of Wiegert and Evans (1964) that annual growth in stable ecosystems should be equal to dead material disappearing in a year.

In the presence of grazing animals growth measurements are even more complicated. These are usually made in areas protected from grazing, but protection affects the micro-environment and the area is no longer under the influence of grazing. Growth is affected by both influences, as will be discussed below.

The effort involved in obtaining statistically acceptable results for growth and utilization in grazed swards may be prohibitive. Waddington and Cooke (1971) found that with less than four replicates, up to 30 caged sites were required in a plot to estimate a 95 percent confidence interval for annual growth or consumption. The variability induced by grazing is a large contributing factor to the need for many sampling sites. The precision of herbage estimates in grazed grasslands has been discussed by many workers, e.g. Green, Langer and Williams (1953) and Wilson (1966).

A very comprehensive review on techniques for measuring herbage growth (' aerial primary production ') was produced by Singh, Lauenroth and Steinhorst (1975). They also tested thirteen procedures at ten sites on six types of grassland over one to three years for grazed and ungrazed situations. The methods ranged from measuring peak standing crop of current live material to the summation of increments in live, recent dead, old dead and litter (trough-peak analysis) with statistical significance constraints. They found large significant differences in estimated growth, but all methods except one were significantly correlated. The difficulty in assessing methods like these is that there is no known true measurement for evaluation. The authors concluded that the most logical choice from a utilitarian and theoretical viewpoint was the trough-peak analysis in all categories of above-ground material with some statistical constraints.

Despite its shortcomings many investigators use simple sequential sampling to estimate growth and utilization as will be discussed below.

1. Growth in ungrazed swards

Annual growth is commonly calculated by the summation of yield differences over periods of a number of days or weeks. Grasslands used only for fodder conservation or green feed are often sampled only at the start of the growing season and whenever the herbage is cut. Annual growth (AG) is then calculated as follows: $AG = \Sigma_1^n (y_n - y_{n-1})$, where y_n is the mean yield of above-ground material of a number of quadrats or strips at the n_{th} sampling period and y_{n-1} that of the previous period.

2. Growth in grazed swards

In the case of rotational or any form of intermittent grazing, growth estimates are made up from periods of grazing and periods of rest. Growth during rest periods is obtained by sampling open quadrats at the end of one grazing period and the beginning of the next. When grazing takes place over a very short period, say one or two days, growth during the grazing period is often ignored. Alternatively, the daily growth rate calculated during the rest period may be used to estimate growth during grazing. However, when grazing takes place over an extended period, growth cannot be ignored and growth rate during the rest period cannot be used to estimate growth during grazing, because growing conditions cannot be assumed to remain constant. In such cases, growth during the grazing period is estimated using quadrats protected from grazing by a cage or movable fence. The objections to this procedure will be discussed in a later section. The sampling technique used is as follows :

A number of sampling sites is selected at random at the beginning of the grazing period. At each site two sampling units, as similar as possible in composition and yield, are marked. On one of these, chosen at random by tossing a coin, the yield is estimated (OY_1) and the other is protected from grazing, and its yield is estimated at the end of a chosen time interval (CY_2). If grazing is continuous, two other sampling units are selected nearby at this time and the process repeated, resulting in yield estimate OY_2. Growth (G_1) can now be calculated for the first period as : $G_1 = CY_2 - OY_1$, for the second period as $G_2 = CY_3 - OY_2$ etc. For annual growth the formula is

$$AG = \Sigma_1^n (CY_n - OY_{n-1}).$$

So-called trimming cuts are sometimes carried out at the end of a grazing period. Although this will reduce variability, it is not recommended except where the practice is part of normal management. Where it is not normal practice this will introduce an artificial factor that may influence productivity. In the case of tropical pastures containing trailing legumes it may result in reduced legume content.

3. Utilization

Utilization in this context means consumption by livestock and other herbivores, decomposition and herbage spoiled by trampling and fouling, because these factors cannot be separated out under normal grazing conditions.

There are two approaches to the estimation of utilization. One is where the main interest lies with the pasture, the other where it mainly concerns the animal.

In the first case, the investigator may wish to determine the relative acceptability or palatability of a range of pasture species. This is experimentally quite simple and can be done by sampling each plot prior to grazing by a large number of animals over a short time and sampling again immediately after grazing. The difference in quantity of vegetation on both occasions may be taken as a measure of acceptability. For simple comparative purposes, acceptability may be estimated by eye using a relative scale from 0 to 10 where zero signified that the plot or species has not been grazed and 10 where it has been grazed to the maximum possible. However, it should be realised that differences in acceptability are influenced by the species included in the comparison, and that such differences do not necessarily reflect intake differences if the animals were confined to separate fields of the individual species.

The approach is quite different when pasture utilization is to be related to animal performance. If grazing is over an extended period the exclosure technique may be used to provide an estimate of herbage utilization, taking into account growth during the interval of study. Using the same symbols as defined above, utilization (U) as calculated according to Linehan, Lowe and Stewart (1947, 1952) is :

$$U = (OY_1 - OY_2) \frac{(\log CY_2 - \log OY_2)}{(\log OY_1 - \log OY_2)}$$

Bosch (1956) simplified this formula to

$$U = OY_1 - OY_2 + \frac{CY_2 - OY_1}{2}$$

after comparing the two on the same data and obtaining practically the same results. This would appear to be a straightforward method of calculating grassland utilization, but many workers have abandoned its use because of the great variability leading to poor results. Only where very intensively used homogeneous fields are involved, can useful results be expected (Linehan, 1952). For further discussions on the methods for the estimation of utilization, see the papers of Linehan, Lowe and Stewart (1947, 1952), Linehan (1952), Carter (1962) and American Society of Range Management and the Agricultural Board (1962).

The reverse use of feeding standards and direct animal intake measurements are discussed in Chapter 7.

4. Effects of exclosures on herbage growth

Many different types of exclosures have been used in grazing studies, ranging from fences to rigid cages and loose panels erected on the spot (Plate 11). The type of exclosure to be used depends on the vegetation, the sampling procedure and the kind of animal to be excluded. Mobility of the exclosures should also

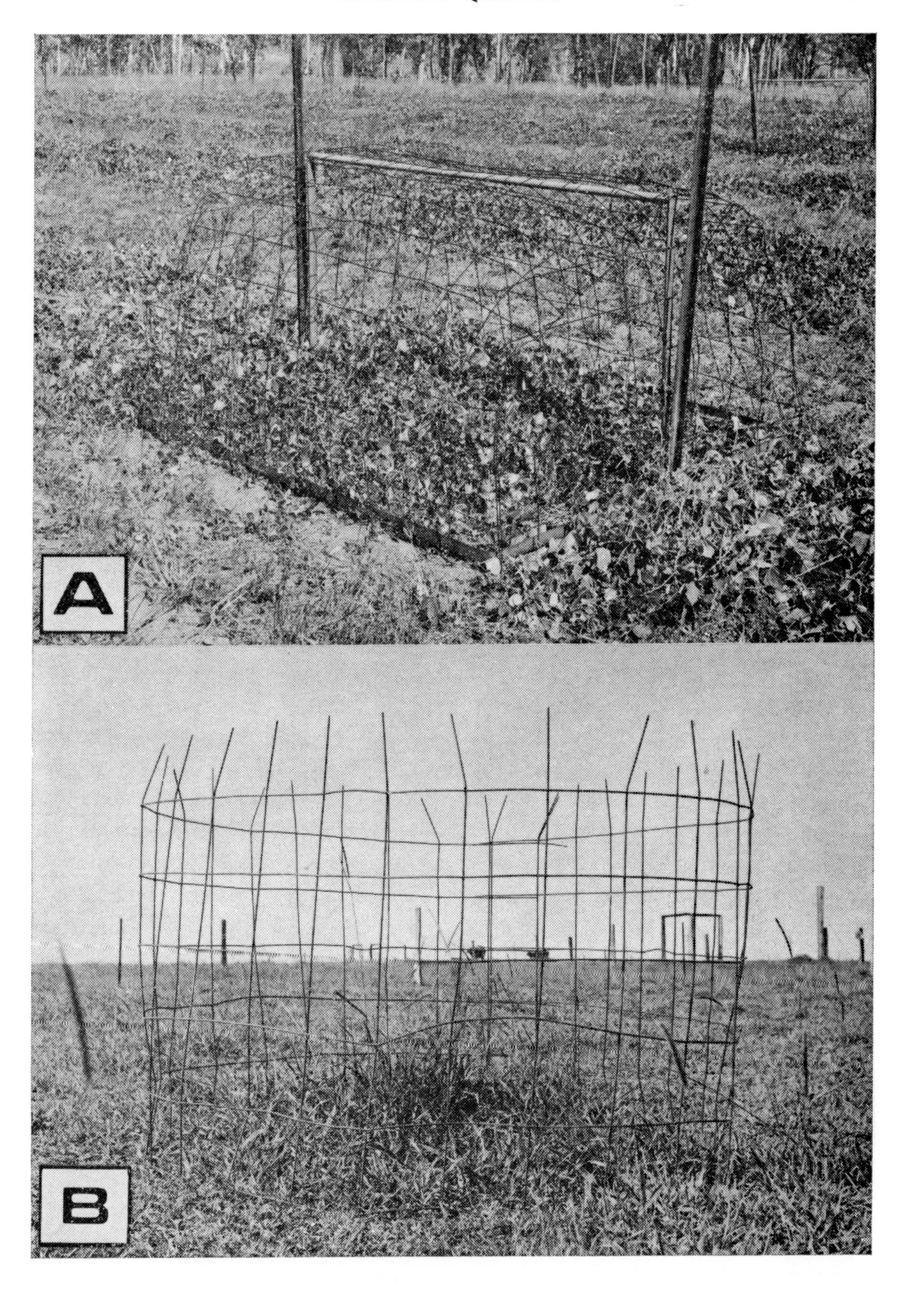

PLATE 11A, B. Two simple cages to exclude grazing. A. Mounted on frame. B. Single sheet of mesh-wire erected in a circle

be considered. Fences provide the least and loose panels the most flexibility. Rigid cages are movable, but bulky. Prendergast and Brady (1955) described a movable cage with electrically charged wires.

The main consideration regarding exclosures in grassland studies is their effect on growth. Cowlishaw (1951) and Owensby (1969) found that growth under cages was greater than outside in the absence of grazing, but Dobb and Elliott (1964) reported reduced growth and reduced seeding of *Festuca rubra* under cages. Heady (1957) recorded higher yields on Californian annual grassland under wire cages (3·75 cm mesh) than outside during the cooler months, but no differences during periods of rapid growth. Jagtenberg and De Boer (1958) in the Netherlands found that yield under cages was 15 percent higher and the herbage had a lower dry matter content than that outside in grasslands on clay soils, but there were no differences on sandy or peaty soils. Grelen (1967) compared ' difference ' methods using stationary cages moved once a year and transient cages, moved monthly, with hand-plucking under a stationary cage every month to the level of the surrounding open range to estimate growth and utilization. Plucking resulted in lower yield and utilization data than yearly or monthly cage measurements. The stationary cage method was considered insufficiently accurate to be recommended for use.

Williams (1951) studied the effect of cages, fences and hurdles on the micro-environment of the enclosed areas. All three resulted in considerable reduction of wind velocity and in an increase in relative humidity inside the structures. Hurdles and fences caused higher temperatures, but cages had a variable response although the temperature inside the cage was usually lower. Cages were also found to reduce light intensity. Dobb and Elliott (1964) largely confirmed these findings and added that soil moisture and soil temperature were not affected, but that vapour-pressure deficit was reduced. These effects on the micro-environment within exclosures can be reduced by using wire with a larger mesh or electrical cages as described by Prendergast and Brady (1955).

Differences due to the absence of grazing animals also add to the artificiality of exclosures. Large differences can be expected between protected and grazed areas due to trampling and fouling. Edmond (1962, 1963) found that treading by sheep reduced yield, apparently mainly because of a reduction in tiller numbers. High levels of soil moisture aggravated the situation. He also reported a change in botanical composition, as some species are more tolerant to treading than others. Smith, Arnott and Peacock (1971) compared clipping and grazing by sheep in the absence of treading and return of dung and urine. The two methods resulted in different sward structures, and in different yields at individual harvests, but not in differences in net regrowth between harvests.

Return of dung and urine to grazing areas results in uneven growth because of the high concentration of nutrients in fouled areas and in uneven grazing because of the unacceptability of fouled herbage. These effects have been reviewed by Sears and Thurston (1953) for pastures grazed by sheep, by Petersen, Lucas and Woodhouse (1956), Norman and Green (1958) and Weeda (1967) for pastures grazed by cattle and by Dirven and De Vries (1973) for horse-pastures.

It is obvious from the above that the magnitude of the effect of exclosures is directly related to the length of time they are in a given position.

V. CONCLUSION AND RECOMMENDATIONS

This review has considered only a selection of available methods, but despite this the investigator is still faced with a multitude to choose from. Main considerations will be the magnitude of the sampling, the purpose for which the data are to be used, manpower and facilities available and the precision required. In conclusion, some commonly encountered situations will be described and recommendations made for suitable methods of yield determination.

1. Small-plot work

Nearly all grassland development or improvement work involves a comparison of new plant material, or of management techniques, and the first stage of their evaluation is usually carried out in small plots measured in m^2. These must be sampled for herbage in order to compare the items under investigation. Where this concerns monocultures, the use of a capacitance meter (Plate 9) may prove the quickest, least demanding method. However, consideration must be given to the need for separate calibrations for different moisture contents in the material which may be brought about by differences in species, or differences in management techniques such as defoliation frequency and height, or fertilizer use. All types of swards may be cut by hand or by one of the many power-scythes or mowers, with or without collection trays or bins. In the case of mixed swards of which the botanical composition needs to be determined, hand cutting and simultaneous sorting is quickest where the components are easily distinguishable, as in open grassland or in grassland consisting of species differing in growth habit. Otherwise, a sub-sample for separation in the laboratory must be taken.

2. Grazing experiments

Grazing experiments are interpreted from animal production measurements, as described in Chapter 7, in combination with pasture measurements, which should be designed to help explain animal performance. Although animal performance commonly cannot be related to the total amount of herbage available except when this is below minimum requirements, total herbage biomass is the base measurement of more meaningful variables, such as total digestible nutrients, green material, amount of legume, etc. It is therefore important that qualifying parameters are measured in addition to total herbage. Another reason for measuring herbage in grazed grassland is to measure the effect of grazing and grazing management on the vegetation, both for total biomass and component species.

The most suitable method will depend on the type of vegetation, the total area to be sampled, the topography and the available facilities. Small grazing experiments usually present few problems and can be sampled in much the same way as small plots. However, the time required for sampling becomes a critical consideration as the area to be sampled increases. Investigators who try to compromise by reducing the number of sampling units, or the total area sampled, are in danger of jeopardising their study. Either sampling efficiency should be improved, for example by the ranked-set method of McIntyre (1952) as described in Chapter 2, or another method should be used. This could either mean more mechanisation, or the use of one of the non-destructive methods discussed in section III : 2 (a). Which approach will be the most suitable depends on factors such as facilities, manpower and topography. Where the land is not accessible by tractor because of steepness or obstructions, a non-destructive method would be the most logical to use. Otherwise, the use of a tractor-driven forage harvester (Plate 7) to harvest strips allows for adequate sampling within a reasonable time.

When sampling for yield, arrangements must be made for botanical and chemical analysis. In the case of non-destructive methods, such as the ' comparative yield ' method of Haydock and Shaw (1975), investigators have an option to include 't Mannetje and Haydock's (1963) ' dry-weight rank ' method of botanical analysis in the same sampling units. In the case of machine harvesting, it is recommended to carry out a separate sampling for botanical analysis; this may either be quadrat harvesting and separation of species, or a visual method (see Chapter 3). The latter is preferred in large experiments because of the time and effort involved in harvesting and sorting a sufficient number of quadrats to satisfy statistical requirements.

Before estimates of growth or utilization are undertaken, careful consideration should be given to the requirements necessary to achieve meaningful results. These may well exceed the available resources, in which case it is unwise to compromise by reducing sampling intensity as this may merely result in obtaining useless data. Sample taking for chemical analysis is described in Chapter 5.

VI. REFERENCES

ALDER, F. E. & RICHARDS, J. A. (1962). A note in the use of the power-driven sheep-shearing head for measuring herbage yields. *J. Br. Grassl. Soc.* **17,** 101-2.

ALEXANDER, C. W., SULLIVAN, J. T. & MCCLOUD, D. E. (1962). A method for estimating forage yields. *Agron. J.* **54,** 468-9.

ALLEN, R. J., CASSELMAN, T. W. & THOMAS, F. H. (1968). An improved forage harvester for experimental plots. *Agron. J.* **60,** 584-5.

AMERICAN SOCIETY OF AGRONOMY in joint committee with Am. Dairy Sci. Assoc., Am. Soc. Anim. Prod. & Am. Soc. Range Mgmt. (1962). Pasture and range research techniques. Ithaca: Comstock Publishing Association.

AMERICAN SOCIETY OF RANGE MANAGEMENT in joint committee with the Agricultural Board (1962). Basic problems and techniques in range research. *Publ. No.* 890, *Natl. Acad. Sci.* Washington, D.C.: Natl. Res. Counc. Publ.

ANSLOW, R. C. (1957). Investigation into the potential productivity of ' Acacia ' (*Leucaena glauca*) in Mauritius. *Rev. Agric. Sucr. Ile Maurice.* **36,** 39-49.

BAKHUIS, J. A. (1960). Estimating pasture production by use of grass length and sward density. *Neth. J. Agric. Sci.* **8,** 211-24.

BOSCH, S. (1956). The determination of pasture yield. *Neth. J. Agric. Sci.* **4,** 305-13.

BOYD, D. A. (1949). Experiments with leys and permanent grass. *J. Br. Grassl. Soc.* **4,** 1-10.

BROWN, Dorothy (1954). Methods of surveying and measuring vegetation. *Commonw. Bur. Pastures Field Crops, Hurley, Bull.* 42.

BRYANT, A M, PARKER, O F., COOK, M. A. S. & TAYLOR, M. J. (1971). An evaluation of the performance of the capacitance meter for estimating the yield of dairy pastures. *Proc. N.Z. Grassl. Assoc.* **33,** 83-9.

BUKER, R. J. (1967). Forage plot harvester. *Agron. J.* **59,** 203-4.

BURROWS, W. H. & BEALE, I. F. (1970). Dimension and production relations of mulga (*Acacia aneura* F Muell.) trees in semi-arid Queensland. *Proc.XIth Int. Grassl Congr.* 33-5.

CAMPBELL, A. G., PHILLIPS, D. S. M. & O'REILLY, E. D. (1962). An electronic instrument. for pasture yield estimation. *J. Br. Grassl. Soc.* **17,** 89-100.

CAMPBELL, N. A. & ARNOLD, G. W. (1973). The visual assessment of pasture yield. *Aust. J. Exp. Agric. Anim. Husb.* **13,** 263-7.

CARPENTER, L. H., WALLMO, O. C. & MORRIS, M. J. (1973). Effect of woody stems on estimating herbage weights with a capacitance meter. *J. Range Mgmt.* **26,** 151-2.

CARTER, J. F. (1962). Herbage sampling for yield: tame pastures. *In* Pasture and range research techniques. By joint committee of Am. Soc. Agron., Am. Dairy Sci. Assoc., Am. Soc. Anim. Prod. & Am. Soc. Range Mgmt., Ithaca: Comstock Publishing Association, pp. 90-101.

COLLINS, K. L., RHYKERD, C. L. & NOLLER, C. H. (1969). A self-propelled experimental plot forage harvester. *Agron. J.* **61,** 338-9.

COOPER, C. S., HYDER, D. N., PETERSEN, R. G. & SNEVA, F. A. (1957). The constituent differential method of estimating species composition in mixed hay. *Agron. J.* **49,** 190-3.

COWLISHAW, S. J. (1951). The effect of sampling cages in the yields of herbage. *J. Br. Grassl. Soc.* **6,** 179-82.

CURRIE, P. O., MORRIS, M. J. & NEAL, D. L. (1973). Uses and capabilities of electronic capacitance instruments for estimating standing herbage. Part 2. Sown ranges. *J. Br. Grassl. Soc.* **28,** 155-60.

DANN, P. R. (1966). A calibration method for estimating pasture yield. *J. Aust. Inst. Agric. Sci.* **32,** 46-9.

DE BOER, T. A. (1956). Een globale grasland-vegetatiekartering van Nederland. *Versl. Landbouwkd. Onderz.* 62-5.

DE VRIES, D. M. (1937). Methods used in scientific plant sociology and in agricultural botanical grassland research. *Herb. Rev.* **5,** 187-93.

DE VRIES, D. M. (1940). De drooggewichtsanalytische methode van botanisch graslandonderzoek voor beweid land. *Versl. Landbouwkd. Onderz.* 46 (1) A. 1-19.

DIRVEN, J. G. P. & DE VRIES, D. M. (1973). Botanische Zusammensetzung von Pferdeweiden. *Z. Acker—U. Pflanzenbau.* **137,** 123-30.

DOBB, J. L. & ELLIOTT, C. R. (1964). Effect of pasture sampling cages on seed and herbage yields of creeping red fescue. *Can. J. Plant Sci.* **44,** 96-9.

EDMOND, D. B. (1962). Effects of treading pasture in summer under different soil moisture levels. *N.Z. J. Agric. Res.* **5,** 389-95.

EDMOND, D. B. (1963). Effects of treading perennial ryegrass (*Lolium perenne* L.) and white clover (*Trifolium repens* L.) pastures in winter and summer at two soil moisture levels. *N.Z. J. Agric. Res.* **6,** 265-76.

EVANS, R. A. & JONES, M. B. (1958). Plant height times ground cover versus clipped samples for estimating forage production. *Agron. J.* **50,** 504-6.

FERRARI, T. J. (1953). The accuracy of yields of grassland and oats evaluated by eye estimates. *Neth. J. Agric. Sci.* **1,** 88-96.

FLETCHER, J. E. & ROBINSON, M. E. (1956). A capacitance meter for estimating forage weight. *J. Range Mgmt.* **9,** 96-7.

FORTMANN, H. R. (1956). Harvester for experimental forage plots. *Agron. J.* **48,** 241-2.

GRANT, S. A. (1971). The measurement of primary production and utilization on heather moors. *J. Br. Grassl. Soc.* **26,** 51-8.

GRASSLAND RESEARCH INSTITUTE, HURLEY (1961). Research techniques in use at the Grassland Research Institute, Hurley. *Commonw. Bur. Pastures Field Crops, Hurley, Berkshire, Bull.* 45.

GREEN, J. O., LANGER, H. J. & WILLIAMS, T. E. (1953). Sources and magnitudes of experimental errors in grazing trials. *Proc. VIth Int. Grassl. Congr.* 1374-9.

GREIG-SMITH, P. (1964). Quantitative plant ecology, London: Butterworths Scientific Publishers, 2nd Edn.

GRELEN, H. E. (1967). Comparison of cage methods for determining utilization of pine-bluestem range. *J. Range Mgmt.* **20,** 94-6.

HAYDOCK, K. P. & SHAW, N. H. (1975). The comparative yield method for estimating dry matter yield of pasture. *Aust. J. Exp. Agric. Anim. Husb.* **15,** 663-70.

HEADY, H. F. (1957). Effect of cages on yield and composition in the California annual type. *J. Range Mgmt.* **10,** 175-7.

HEBBLETHWAITE, P. & HUGHES, M. (1959). Estimating the soil content of herbage collected by forage harvester. *J. Br. Grassl. Soc.* **14,** 169-71.

HEDRICK, D. W. & HITCHCOCK, G. (1953). Use of scythette in range forage studies. *J. Range Mgmt.* **6,** 182-4.

HORMAY, A. L. (1943). A method of estimating grazing use of bitterbrush. *U.S. Dep. Agric., For. Serv., Calif. For. Range Exp. Sta. Res. Note* 35.

HOWELL, H. B. (1956). A new experimental plot harvester. *Agron. J.* **48,** 240-1.

HUGHES, E. E. (1962). Estimating herbage production using inclined point frame. *J. Range Mgmt.* **15,** 323-5.

HUNTER, R. F. & GRANT, S. A. (1961). The estimation of ' green dry matter ' in a herbage sample by methanol-soluble pigments. *J. Br. Grassl. Soc.* **16,** 43-5.

HURD, R. M. (1959). Factors influencing herbage weight of Idaho fescue plants. *J. Range Mgmt.* **12,** 61-3.

HUTCHINGS, S. S. & SCHMAUTZ, J. E. (1969). A field test of the relative-weight-estimate method for determining herbage production. *J. Range Mgmt.* **22,** 408-11.

HUTCHINSON, K. J. (1967). A coring technique for the measurement of pasture of low availability to sheep. *J. Br. Grassl. Soc.* **22,** 131-4.

HUTCHINSON, K. J., MCLEAN, R. W. & HAMILTON, B. A. (1972). The visual estimation of pasture availability using standard pasture cores. *J. Br. Grassl. Soc.* **27,** 29-34.

HUTTON, E. M. & BONNER, I. A. (1960). Dry matter and protein yields in four strains of *Leucaena glauca* Benth. *J. Aust. Inst. Agric. Sci.* **26,** 276-7.

JAGTENBERG, W. D. (1970). Predicting the best time to apply nitrogen to grassland in spring. *J. Br. Grassl. Soc.* **25,** 266-71.

JAGTENBERG, W. D. & DE BOER, T. A. (1958). De invloed van graskooien op de grasopbrengst. *Landbouwk. Tijdschr* **70,** 879-89.

JOHNS, G. G. (1972). The accuracy of a range of capacitance probe methods for estimating pasture yields. *J. Agric. Sci.* **79,** 273-80.

JONES, R. K. (1973). Cutting pasture samples with a new type of shearing handpiece. *Trop. Grassl.* **7,** 251-3.

JONES, R. M. & SANDLAND, R. L. (1974). Use of an electronic meter to estimate pasture yield. *Aust. CSIRO, Div. Trop. Agron. Ann. Rep.* 1973-1974, p. 54.

JONES, R. J. & HAYDOCK, K. P. (1970). Yield estimation of tropical and temperate pasture species using an electronic capacitance meter. *J. Agric. Sci.* **75,** 27-36.

KEMP, J. G. & KALBFLEISCH, W. (1957). A crop harvester for forage plots. *Can. J. Plant Sci.* **37,** 418-22.

KENNEDY, R. K. (1972). The sickledrat: a circular quadrat modification useful in grassland studies. *J. Range Mgmt.* **25,** 312-3.

LINEHAN, P. A. (1952). Use of cage and mower-strip methods for measuring the forages consumed by grazing animals. *Proc. VIth. Int. Grassl. Cong.* 1328-33.

LINEHAN, P. A., LOWE, J. & STEWART, R. H. (1947). The output of pasture and its measurement. Part II. *J. Br. Grassl. Soc.* **2,** 145-68.

LINEHAN, P. A., LOWE, J. & STEWART, R. H. (1952). The output of pasture and its measurement. Part III. *J. Br. Grassl. Soc.* **7,** 73-98.

LOMNICKI, A., BANDOLA, E. & JANKOWSKA (1968). Modification of the Wiegert-Evans method for estimation of net primary production. *Ecology* **49,** 147-9.

MCGINNIES, W. J. (1959). A rotary lawn mower for sampling range herbage. *J. Range Mgmt.* **12,** 203-4.

MCINTYRE, G. A. (1952). A method for unbiased selective sampling, using ranked sets. *Aust. J. Agric. Res.* **3,** 385-90.

MANNETJE, L. 't & HAYDOCK, K. P. (1963). The dry-weight-rank method for the botanical analysis of pasture. *J. Br. Grassl. Soc.* **18,** 268-75.

MARTIN, T. J. & TORSSELL, B. W. R. (1971). A self-propelled harvester for pasture crops. *J. Aust. Inst. Agric. Sci.* **37,** 244-6.

MATCHES, A. G. (1963). A cordless hedge trimmer for herbage sampling. *Agron. J.* **55,** 309.

MATCHES, A. G. (1966). Sample size for mower-strip sampling of pastures. *Agron. J.* **58,** 213-5.

MILNER, C. & HUGHES, R. E. (1968). Methods for the measurement of the primary production of grassland. Int. Biol. Program, Handbook 6. Oxford: Blackwell Scientific Publications.

MITCHELL, J. E. (1972). An analysis of the beta-attenuation technique for estimating standing crop of prairie range. *J. Range Mgmt.* **25,** 300-4.

MITCHELL, K. J. & GLENDAY, A. C. (1958). The tiller population of pastures. *N.Z. J. Agric. Res.* **1,** 305-18.

MORLEY, F. H. W., BENNETT, D. & CLARK, K. W. (1964). The estimation of pasture yield in large grazing experiments. *Aust. CSIRO Div. Plant Ind. Field Stn. Rec.* **3,** (2), 43-7.

MORRIS, M. J. (1967). An abstract bibliography of statistical methods in grassland research. *US Dep. Agric. For. Serv. Misc. Publ. No.* 1030.

MOTT, G. O., BARNES, R. F. & RHYKERD, C. L. (1965). Estimating pasture yield *in situ* by beta ray attenuation techniques. *Agron. J.* **57,** 512-3.

NEAL, D. L. & NEAL, J. L. (1973). Uses and capabilities of electronic capacitance instruments for estimating standing herbage. Part 1. History and development. *J. Br. Grassl. Soc.* **28,** 81-9.

NORMAN, M. J. T. & GREEN, J. O. (1958). The local influence of cattle dung and urine upon the yield and botanical composition of permanent pasture. *J. Br. Grassl. Soc.* **13,** 39-45.

OWENSBY, C. E. (1969). Effect of cages on herbage yield in true prairie vegetation. *J. Range Mgmt.* **22,** 131-2.

PASTO, J. K., ALLISON, J. R. & WASHKO, J. B. (1957). Ground cover and height of sward as a means of estimating pasture production. *Agron. J.* **49,** 407-9.

PATTERSON, J. J. & BROWNING, D. R. (1962). Hydraulic self-propelled forage plot harvester. *Agric. Eng.* **43,** 270-1, 291.

PEARSON, R. L. & MILLER, L. D. (1973). The biometer—a hand-hold grassland biomass meter. *U.S.A. Colorado State Univ. (Fort Collins), Dept. Watershed Sci. Incidental Rep. No.* 6.

PECHANEC, J. P. & PICKFORD, G. D. (1937). A weight estimate method for the determination of range or pasture production. *J. Am. Soc. Agron.* **29,** 894-904.

PETERSEN, R. G., LUCAS, H. L. & WOODHOUSE, W. W. (1956). The distribution of excreta by freely grazing cattle and its effect on pasture fertility: *Agron. J.* **48,** 440-9.

PHILLIPS, D. S. M. & CLARKE, S. E. (1971). The calibration of a weighted disc against pasture dry matter yield. *Proc. N.Z. Grassl. Assoc.* **33,** 68-75.

PRENDERGAST, J. J. & BRADY, J. J. (1955). Improved movable cage for use in grassland research. *J. Br. Grassl. Soc.* **10,** 189-90.

SCOTT, D., MENALDA, P. H. & BROUGHAM, R. W. (1968). Spectral analysis of radiation transmitted and reflected by different vegetations. *N.Z. J. Bot.* **6,** 427-49.

SEARS, P. D. (1951). The technique of pasture measurement. *N.Z. J. Sci. Technol. Sect. A.,* **33,** 1-29.

SEARS, P. D. & THURSTON, W. G. (1953). Effect of sheep droppings on yield, botanical composition and chemical composition of pasture. III. Results of field trial at Lincoln, Canterbury, for the years 1944-1947. *N.Z. J. Sci. Technol., Sect. A.,* **34,** 445-59.

SINGH, J. S., LAUENROTH, W. & STEINHORST, R. K. (1975). Review and assessment of various techniques for estimating net aerial primary production in grasslands from harvest data. *Bot. Rev.* **41,** 181-232.

SMITH, A., ARNOTT, R. A. & PEACOCK, J. M. (1971). A comparison of the growth of a cut sward with that of grazed swards, using a technique to eliminate fouling and treading. *J. Br. Grassl. Soc.* **26,** 157-62.

SPEDDING, C. R. W. & LARGE, R. V. (1957). A point-quadrat method for the description of pasture in terms of height and density. *J. Br. Grassl. Sovc.* **12,** 229-34.

SWALLOW, C. (1967). Self-propelled plot forage harvester. *Agron. J.* **59,** 609-10.

SYMONS, L. B. & JONES, R. I. (1971). An analysis of available techniques for estimating production of pastures without clipping. *Proc. Grassl. Soc. S. Afr.* **6,** 185-90.

TAKAHASHI, M. & RIPPERTON, J. C. (1949). Koa haole (*Leucaena glauca*): Its establishment, culture, and utilization as a forage crop. *Univ. Hawaii Agric. Exp. Stn. Bull.* 100.

TEARE, I. D. (1963). Estimating forage yield *in situ. Thesis for Ph. D., Purdue Univ., USA.*

TEARE, I. D., MOTT, G. O. & EATON, J. R. (1966). Beta attenuation—a technique for estimating forage yield *in situ. Radiat. Bot.* **6,** 7-11.

THOMPSON, A & RAVEN, A. M. (1955). Soil contamination of herbage samples. *J. Sci. Food Agric.* **6,** 768-77.

THOMPSON, J. L. & HEINRICHS, D. H. (1963). Note on the Swift Current Forage Plot Harvester II. *Can. J. Plant Sci.* **43,** 602-4.

TIWARI, D. K., JACKOBS, J. A. & CARMER, S. G. (1963). Statistical technique for correcting botanical or floristic estimates in pasture research. *Agron. J.* **55,** 226-8.

VAN DYNE, G. M. (1966). Use of a vacuum-clipper for harvesting herbage. *Ecology* **47,** 624-6.

VAN DYNE, G. M., VOGEL, W. G. & FISSER, H. G. (1963). Influence of small plot size and shape on range herbage production estimates. *Ecology* **44,** 746-59.

WADDINGTON, J. & COOK, D. A. (1971). The influence of sample size and number on the precision of estimates of herbage production and consumption in two grazing experiments. *J. Br. Grassl. Soc.* **26,** 95-101.

WALLENTINUS, H. G. (1973). Above-ground primary production of a *Juncetum gerardi* on a Baltic sea-shore meadow. *Oikos* **24,** 200-19.

WEEDA, W. C. (1967). The effect of cattle dung patches on pasture growth, botanical composition, and pasture utilisation. *N.Z. J. Agric. Res.* **10,** 150-9.

WELLS, K. F. (1971). Measuring vegetation changes on fixed quadrats by vertical ground stereophotography. *J. Range Mgmt.* **24,** 233-6.

WIEGERT, R. G. & EVANS, F. C. (1964). Primary production and the disappearance of dead vegetation on an old field in Southeastern Michigan. *Ecology* **45,** 49-63.

WILLIAMS, S. S. (1951). Microenvironment in relation to experimental techniques. *J. Br. Grassl. Soc.* **6,** 207-17.

WILM H. G., COSTELLO, D. F. & KLIPPLE, G. E. (1944). Estimating forage yields by the double-sampling method. *J. Am. Soc. Agron.* **36,** 194-203.

WILSON, D. B. (1966). Variability in herbage yields from caged areas in a pasture experiment. *Can. J. Plant Sci.* **46,** 249-55.

WHITNEY, A. S. (1974). Measurement of foliage height and its relationships to yields of two tropical forage grasses. *Agron. J.* **66,** 334-6.

WOODHOUSE, W. W., PETERSON, R. G. & BERENYI, N. (1963). Hatteras Sampler. *Agron. J.* **55,** 410-1.

Chapter 5

SAMPLE PREPARATION AND CHEMICAL ANALYSIS OF VEGETATION

A. D. Johnson

I. INTRODUCTION

Most studies involving herbage production require some form of chemical or biological qualification of the dry matter. The choice of elements to be determined, or essays to be carried out, and the method of analytical procedure to be used depend on the objective of the research, available equipment, and the desired precision. Three general types of investigation may be recognised, each with different analytical requirements :

(i) Studies which require very precise analyses, e.g. plant nutrition.

(ii) Survey investigations such as in plant introduction or breeding, in which a broad spectrum analysis, but usually not high precision, is called for.

(iii) Agronomic field experiments, in which various parameters are to be monitored with reasonable accuracy over a period of time.

The actual analyses are usually carried out under the direction of analytical chemists, but the samples are collected, handled, dried and ground by agronomists. The latter must be aware that his handling of the samples will influence the accuracy of the results. Some of these problems will be discussed here and the methods of analysis used in this laboratory will be mentioned briefly. Some other publications that may be consulted on sample preparation are Grassland Research Institute, Hurley (1961), and Milner and Hughes (1968). Chemical procedures are fully described in various textbooks.

II. SAMPLING

As chemical composition varies with plant species, plant part, physiological age of the whole plant and of individual plant parts, the sampling procedures used must ensure that the material submitted for analysis is representative of that under study. In general, mineral concentrations and digestibility are higher and fibre content lower in young material, and higher in leaves than in old material and stems. Therefore, it depends on the objective of the investigation as to what constitutes an appropriate sample. In studies designed to monitor the concentration of one or more elements over a period of time, it is desirable to collect plant parts of equivalent physiological age, for example the last fully expanded leaf in the case of grasses. However, when studying mineral uptake by plants, whole-plant material should be analysed, although underground parts are usually omitted. In animal diet studies, it is important that material similar to that actually eaten is collected for analysis. In such cases whole-plant samples would underestimate the concentration of minerals and the digestibility of material ingested, because animals eat selectively. Representative samples of what grazing animals eat may be obtained by hand-plucking or from oesophageal fistulae (see Chapter 7).

Contamination by soil, fertilizers or animal excreta causes errors on two accounts, firstly through direct enrichment and secondly through weight errors. The latter can be overcome by ashing in order to express the chemical composition on an organic matter basis. Surface contamination is sometimes eliminated by washing with water, but this may lead to substantial losses of dry matter and major and minor elements (Thompson & Raven, 1955; Long, Sweet & Tukey, 1956; Hebblethwaite & Hughes, 1959; Bremner & Knight, 1970). In the case of very critical analysis, brushing individual leaves may be the safest method of reducing surface contamination. Contamination of ocsophageal fistula samples by saliva is discussed in Chapter 7.

III. DRYING

The objective of drying samples before analysis is to remove moisture with minimum loss of dry matter, chemical constituents or feeding value. Dry matter losses are a result of respiration of carbohydrate substrates, and losses of between 2 and 17 percent of the total dry matter have been reported (McRostie & Hamilton, 1927; Raymond & Harris, 1954; Greenhill, 1960; Minson & Lancaster, 1963). Smith (1973), reviewing the influence of drying and storage conditions on carbohydrate content, pointed out that temperatures over 80°C can cause thermo-chemical degradation, while slow drying at temperatures below 50°C causes dry matter losses by respiration and enzymatic conversions. He concluded that the most acceptable heat-drying results had been obtained by drying for a short time at 100°C, followed by 70°C. However, in practice the effectiveness of this depends on how quickly the whole sample reaches such a high temperature. This is influenced by type of oven, size of sample, bulk density of sample and type of sample container (Wolf & Ellmore, 1975).

Volatile nitrogen losses of 3 to 10 percent have been widely reported (e.g. Pucher, Vickery & Leavenworth, 1935; Pirie, 1956; Sharkey, 1970). Nitrogen loss is accentuated by prolonged wilting of plant tissue, producing extensive proteolysis with resultant formation of amides which are subject to volatilisation at temperatures around 100°C. Drying at elevated temperatures can destroy or modify many low molecular weight plant constituents such as amino acids. If these compounds are to be determined, freeze-drying is to be preferred (Mayland, 1968; Burns, Noller & Rhykerd, 1966), but the technique does not lend itself to the drying of large samples unless a massive, high cost installation is used.

The use of drying temperatures over 100°C for extended periods should be avoided when *in vitro* digestibility (Tilley & Terry, 1963) or fibre constituents, particularly lignin (Van Soest, 1965) are to be determined. Cochrane and Brown (1974) found that oven-drying at 60°, 80° or 100°C for 24 h reduced dry matter digestibility of ground, frozen samples from 71 percent down to 67-68 percent, which they ascribed to the possible formation of indigestible compounds (see also Clark & Mott, 1960; Noller *et al.*, 1968).

The use of micro-wave drying ovens has been reported to overcome some of the problems mentioned (Merridew & Raymond, 1953; Carlier & van Hee, 1971). Work being carried out in this laboratory suggests that the technique is efficient for small samples but does not lend itself to large-scale drying.

Although rapid drying as soon as possible after harvesting is recommended, this can seldom be done in practice. The problem is accentuated where the sampling locality is remote from drying facilities. In this connection, Cochrane and Brown (1974) investigated the effect of storage before drying at a range of temperatures on chemical composition and *in vitro* digestibility. They found that storage for 24 h at 21°C had little effect on dry matter (3 percent loss), crude protein or *in vitro* dry matter digestibility, but soluble carbohydrate content was lowered by about 8 percent.

IV. GRINDING

Dry plant samples should be ground in standard laboratory mills fitted with stainless steel sieves (1-mm mesh). In general, mills fitted with brass sieves are unsuitable for minor element work as considerable contamination by copper, zinc, iron and manganese will occur. Samples for special analytical procedures such as emission spectroscopy and X-ray analysis require further grinding in special mills to ensure uniform particle size. It should be borne in mind that every grinding operation results in contamination by one or more elements, depending on the type of mill used (Thompson & Bankston, 1970). Samples for digestibility and fibre analysis can be ground in most laboratory mills, but as particle size is critical, a standard screen size (usually 0·8 – 1·0 mm) is used (McLeod & Minson, 1969). Although the presence of moisture in the sample during grinding has no significant effect upon digestibility and fibre content, grinding is slow and difficult with moist samples and grinding of hot samples is preferred. To overcome problems associated with drying, samples can be ground in a frozen state using liquid nitrogen and dry ice (Brown & Radcliffe, 1970). After grinding, the material should be stored in glass or plastic bottles with close fitting lids.

V. METHODS OF ANALYSIS

Analytical techniques used in pasture sample analysis usually rely on automation and instrumentation. In this laboratory four techniques for chemical analysis are used : autoanalysis, flame emission spectrophotometry, atomic absorption spectrophotometry and, more recently, direct reading emission spectrophotometry. Atomic absorption and autoanalysis techniques produce better precision than direct reading emission techniques, but the latter has the outstanding advantage of multi-element analysis and high speed.

Following Kjeldahl digestion of the pasture samples, nitrogen and phosphorus are determined simultaneously by autoanalysis by a modification of the method of Williams and Twine (1967). Sodium and potassium are determined simultaneously by automatic flame emission spectrophotometry of plant ash solution by the method N-20A of the Technicon Laboratory (Technicon Instruments Corporation, New York). Calcium, magnesium, zinc, manganese and iron are determined by atomic absorption spectrophotometry by a modification of the method of David (1958). Copper is also determined by the same technique by the method of Allan (1961). Rapid multi-element analysis for 21 elements, including nitrogen and sulphur, is carried out by direct reading emission spectrophotometry (Plate 12) by the method of Johnson and Simons (1972).

Several laboratory methods are available for predicting *in vivo* digestibility. The two-stage *in vitro* technique (Clark, 1958; Clark & Mott, 1960; Tilley & Terry, 1963) is the most popular because of its high accuracy, but it has limitations in that a ready source of rumen fluid is required and the process is slow and tedious, with little chance of being completely automated. At this laboratory the method has been modified for the processing of large numbers of samples (Minson & McLeod, 1972).

PLATE 12. Emission spectrometer (quantometer) for simultaneous read-out of 24 elements in herbage samples (Johnson & Simons, 1972)

Although chemical methods such as crude fibre, acid detergent fibre and lignin can be used when *in vitro* laboratory facilities are not available, these are usually not less accurate in predicting digestibility than the *in vitro* technique (Minson, 1971; McLeod & Minson, 1971, 1972, 1974). The most promising of the chemical techniques appears to be the lignin technique of Christian (1971).

VI. REFERENCES

ALLAN, J. E. (1961). The determination of copper by atomic absorption spectrophotometry. *Spectrochim. Acta* **17,** 459-66.

BREMNER, I. & KNIGHT, A. H. (1970). The complexes of zinc, copper and manganese present in ryegrass. *Br. J. Nutr.* **24,** 279-89.

BROWN, D. C. & RADCLIFFE, J. C. (1970). A new method of preparing ground silage for the determination of chemical composition and *in vitro* digestibility. *Proc. XIth Int. Grassl. Congr.*, pp. 750-4.

BURNS, J. C., NOLLER, C. H. & RHYKERD, C. L. (1966). Influence of drying method and fertility treatments on the total and water soluble nitrogen contents of alfalfa. *Agron. J.* **58,** 13-6.

CARLIER, L. A. & VAN HEE, L. P. (1971). Microwave drying of lucerne and grass samples. *J. Sci. Food Agric.* **22,** 306-7.

CHRISTIAN, K. R. (1971). Detergent method for total lignin in herbage. *Aust. CSIRO Div. Plant Ind. Field Stn. Rec.* **10,** 29-34.

CLARK, K. W. (1958). The adaption of an artificial rumen technique to the estimation of the gross digestible energy of forages. *Diss. Abstr.* **19,** 926.

CLARK, K. W. & MOTT, G. O. (1960). The dry matter digestion *in vitro* of forage crops. *Can. J. Plant Sci.* **40,** 123-9.

COCHRAN, M. J. & BROWN, D. C. (1974). Effect of storage and preparation of herbage on chemical composition and *in vitro* digestibility. *J. Aust. Inst. Agric. Sci.* **40,** 67-9.

DAVID, D. J. (1958). Determination of zinc and other elements in plants by atomic-absorption spectroscopy. *The Analyst* **83,** 655-61.

GRASSLAND RESEARCH INSTITUTE, HURLEY (1961). Research techniques in use at the Grassland Research Institute, Hurley. *Commonw. Bur. Pastures Field Crops, Hurley, Berkshire, Bull.* 45.

GREENHILL, W. L. (1960). Determination of the dry weight of herbage by drying methods. *J. Br. Grassl. Soc.* **15,** 48-54.

HEBBLETHWAITE, P. & HUGHES, M. (1959). Estimating the soil content of herbage collected by forage harvester. *J. Br. Grassl. Soc.* **14,** 169-71.

JOHNSON, A. D. & SIMONS, J. G. (1972). Direct reading emission spectroscopic analysis of plant tissue using a briquetting technique. *Commun. Soil Sci. Plant Anal.* **3,** 1-9.

LONG, W. G., SWEET, D. V. & TUKEY, H. B. (1956). The loss of nutrients by leaching of the foliage. *Mich. Agric. Exp. Stn. Q. Bull.* **38,** 528-32.

McLEOD, M. N. & MINSON, D. J. (1969). Sources of variation in the *in vitro* digestibility of tropical grasses. *J. Br. Grassl. Soc.* **24,** 244-9.

McLEOD, M. N. & MINSON, D. J. (1971). The error in predicting pasture dry-matter digestibility from four different methods of analysis for lignin. *J. Br. Grassl. Soc.* **26,** 251-6.

McLEOD, M. N. & MINSON, D. J. (1972). The effect of method of determination of acid-detergent fibre on its relationship with the digestibility of grasses. *J. Br. Grassl. Soc.* **27,** 23-7.

McLEOD, M. N. & MINSON, D. J. (1974). The accuracy of predicting dry matter digestibility of grasses from lignin analysis by three different methods. *J. Sci. Food Agric.* **25,** 907-11.

McROSTIE, G. P. & HAMILTON, R. I. (1927). The accurate determination of dry matter in forage crops. *J. Am. Soc. Agron.* **19,** 243-51.

MAYLAND, H. F. (1968). Effect of drying methods on losses of carbon, nitrogen and dry matter from alfalfa. *Agron. J.* **60,** 658-9.

MERRIDEW, J. N. & RAYMOND, W. F. (1953). Laboratory drying of herbage by radio frequency dielectric heating. *Br. J. Appl. Phys.* **4,** 37-9.

MILNER, C. & HUGHES, R. E. (1968). Methods for the measurement of the primary production of grassland. Int. Biol. Program Handbook 6. Oxford: Blackwell Scientific Publications.

MINSON, D. J. (1971). The place of chemistry in pasture evaluation. *Proc. R. Aust. Chem. Inst.* **38,** 141-5.

MINSON, D. J. & LANCASTER, R. J. (1963). The effect of oven temperature on the error in estimating the dry matter content of silage. *N.Z. J. Agric. Res.* **6,** 140-6.

MINSON, D. J. & McLEOD, M. N. (1972). The *in vitro* technique: its modification for estimating digestibility of large numbers of tropical pasture samples. *Aust. CSIRO Div. Trop. Past. Tech. Pap. No.* 8.

NOLLER, C. H., PRESTES, P. J., RHYKERD, C. L., RUMSEY, T. S. & BURNS, J. C. (1968). Changes in chemical composition and digestibility of forages with method of sample handling and drying. *Proc. Xth Int. Grassl. Congr.*, pp. 429-34.

PIRIE, N. W. (1956). General methods for separation. Making and handling extracts. *In* Modern methods of plant analysis, Eds. K. Paech & M. V. Tracey. Vol. 1. pp. 26-55. Berlin: Springer-Verlag.

PUCHER, G. W., VICKERY, H. B. & LEAVENWORTH, C. S. (1935). Determination of ammonia and of amide nitrogen in plant tissue. *Ind. Eng. Chem. Anal. Ed.* **7,** 152-6.

RAYMOND, W. F. & HARRIS, C. E. (1954). The laboratory drying of herbage and faeces, and dry matter losses possible during drying. *J. Br. Grassl. Soc.* **9,** 119-30.

SHARKEY, M. J. (1970). Errors in measuring nitrogen and dry-matter content of plant and faeces material. *J. Br. Grassl. Soc.* **25,** 289-94.

SMITH, D. (1973). Influence of drying and storage conditions on nonstructural carbohydrate analysis of herbage tissue—a review. *J. Br. Grassl. Soc.* **28,** 129-34.

THOMPSON, A. & RAVEN, A. M. (1955). Soil contamination of herbage samples. *J. Sci. Food Agric.* **6,** 768-77.

THOMPSON, G. & BANKSTON, D. C. (1970). Sample contamination from grinding and sieving determined by emission spectrometry. *Appl. Spectros.* **24,** 210-9.

TILLEY, J. M. A. & TERRY, R. A. (1963). A two-stage technique for the *in vitro* digestion of forage crops. *J. Br. Grassl. Soc.* **18,** 104-11.

VAN SOEST, P. J. (1965). Use of detergents in analysis of fibrous feeds. III. Study of effects of heating and drying on yield of fibre and lignin in forages. *J. Assoc. Off. Agric. Chem.* **48,** 785-90.

WILLIAMS, C. H. & TWINE, J. R. (1967). Determination of nitrogen, sulphur, phosphorus, potassium, sodium, calcium and magnesium in plant material by automatic analysis. *Aust. CSIRO Div. Plant. Ind. Tech. Pap. No.* 24.

WOLF, D. D. & ELLMORE, T. L. (1975). Oven drying of small herbage samples. *Agron. J.* **67** 571-4.

Chapter 6

ANIMAL PRODUCTION STUDIES ON GRASSLAND

F. H. W. Morley

I. INTRODUCTION

Although grasslands are used for recreation, conservation and water supplies, their prime use is for animal production, and this aspect only is to be considered here. Pastures and the animals grazing on them form production systems having inputs such as land, animals, labour and fertilizer. The outputs are animal products. The objectives may be manifold, but generally include :

(i) maximum (or at least substantial) efficiency of production in terms of profit, fulfilment of needs, or other criteria,
(ii) conservation of resources, including land,
(iii) minimum animal stress.

A study of pasture-animal production must, if it is to be even moderately complete, take some account of these objectives and provide the data to estimate them.

An examination of the literature discloses increasing divergence between the objectives of different workers in measuring animal production from grasslands. Some are primarily concerned with whole systems of production. Others, perhaps deterred by the difficulties of the holistic approach, limit their measurements to processes which may not always be relevant to any actual systems of production. The evaluation of grasslands through animal production must generally involve compromises between the desirability of studying a total system on an adequate scale, and the resources available. The relevance of the results must be judged by whether the situation studied is part of an actual system or a potential production system. Unless evaluations are carried out with reference to actual systems they risk being irrelevant; unfortunately, many investigators appear to have failed to appreciate this need.

These viewpoints have been emphasised by Morley and Spedding (1968) and Morley (1972). An increasing number of animal/pasture, soil/plant/animal, weather/soil/plant/animal, and other systems of varying levels of completeness and complexity are being described. The definition of such systems can provide frameworks which enable research to be directed increasingly to evaluating the critical components of relevant systems. These developments could have far-reaching implications for this area of research.

Obviously, if animal production is to be used as a measure of grassland productivity, a vast array of techniques for measurement of animal production is needed. A mastery of appropriate techniques is a prerequisite for measuring animal production, but an adequate description of these would be inappropriate and impractical here. There are many papers, handbooks and other texts on such techniques, written for specialists as well as students, which may be consulted. (See Chapter 7 and the general reading list at the end of this Chapter.)

This Chapter is therefore less concerned with what work has been done or the techniques which have been used than with the concepts which are thought to underlie the application of techniques, and the use of the information gained to increase the efficiency of grassland management and other resources in satisfying human needs. Neither a general review, nor an exhaustive discussion of systems of production is attempted. The objective is to suggest a sufficient number of concepts, illustrated by a few examples, in the hope that the principles which emerge may be applied to increase the relevance of designs, the performance, and the interpretation of measurements of grassland productivity through animal production.

II. GRAZING SYSTEMS, DESCRIPTIONS AND DEFINITIONS

The literature on grassland productivity abounds with terms such as ' rotational grazing ', ' deferred grazing ' and ' stocking rate ' but the precise meaning of many such terms is often unclear. Heady (1970) discussed this confusion and defined a number of terms as follows :

" *Grazing season* is that portion of the year during which grazing is feasible. It may be the whole year or a very short time span and is normally a function of uncontrolled environment such as climate. In this context the vegetative growing season is only a part of the grazing season.

Grazing period is that portion of the grazing season during which grazing takes place. The beginning and end of the grazing period on each land unit are stipulated by the grazing system.

Continuous grazing is defined as unrestricted livestock access to any part of the range throughout a grazing period which encompasses the whole of the grazing season. It may be year-long or shorter depending upon environmental or other restrictions of the range area to grazing livestock. Continuous grazing is principally distinguished from other types on the basis that grazing occurs through the period when forage plants are growing.

Ungrazed signifies a brief period of non-use that is not scheduled specifically to allow seed maturation or seedling establishment. This term permits specific definitions and elimination of double meanings for deferred and rest.

Deferred grazing specifies that the vegetation is not grazed until seed maturity is nearly complete or assured and that it is grazed after seed maturity. Deferment in the second year permits establishment of seedlings. For various reasons many ranchers in effect defer some of their range each year, often the same piece of their holdings.

Rest as a term in range grazing systems has come or is coming to mean that a pasture is not grazed at all in a given year. Even the mature forage is not harvested.

Rotation means that animals are moved from one pasture to another on a scheduled basis. When the rotation is short a pasture may be alternatively grazed and ungrazed several times during a grazing season. If the rotation is long, it may be on a basis of rotating the deferred pasture or the rested pasture among years."

Many may not accept Heady's definitions but they provide a working basis which is generally available for reference. At least any use in a different sense should be clearly defined.

The animal production systems should be sufficiently described, and the terms used related to that description, so that misunderstanding is minimised. The description of the system should include, in addition to the animals and pastures, the number of subdivisions and the rules for moving livestock from one subdivision to another. Systems of ' rotational grazing ' require further definition because the outcome of the system may very greatly depend on the number of pastures and the lengths of periods of, and between, grazings.

The loose use of some terms may readily generate confusion. For example ' stocking rate ' should be a statement of the number of animals on a given area. If the system in question includes many subdivisions it would be preferable to define stocking rate as the number of animals per unit area of the total system, not for a particular subdivision at a particular time. The use of put-and-take systems (see later) creates problems in relation to this term.

' Carrying capacity ' is presumed to relate to the number of animals which can be carried per unit area. It must obviously be so influenced by the kind of production (fattening, breeding, maintenance) and the management employed (supplementary feeding, timing of breeding, marketing, etc.) that the term is of limited value.

The term ' optimum stocking rate ' is being used increasingly widely and often loosely. Obviously, an optimum must relate to some objective, which will almost invariably include economic items. The association of optimum with maximum physical production per unit area seems difficult to justify on any logical basis. This use should therefore be avoided.

Finally, the literature is encumbered with terms such as ' efficiency of pasture production ', which have created confusion and distorted objectives. Investigators should avoid using them, and they should be replaced by precisely defined terms which do not imply values. For example ' pasture wasted ' might be replaced by ' pasture not eaten ' which is less likely to have emotive overtones.

III. THE NEED FOR EVALUATION BY ANIMAL PRODUCTION

Measurements of animal production from grasslands involve substantially more resources than are needed for measurements of vegetation. Therefore, evaluation of grasses and grassland management has tended to be limited to the effects of animals on plants, perhaps supplemented by data on digestibility, diet selection, and even intake. Estimates of animal production from the system must depend on extrapolation from such information.

As investigations advance beyond studies of field botany through spaced plants in the field and in controlled environments, the grazing animal plays an increasingly important role. The objectives of a particular experiment will determine which measurements of the grazing system are appropriate and necessary. These objectives may be oriented towards the soil, the pasture, the animals, or a combination of these. It is conceivable that within the framework of a conceptual grazing system an experiment is designed to investigate the effect of fertilizer or soil fertility on pasture production. Measurements on only soil and pasture may be justified at this stage, but animals should be used to graze the pastures (Brockman, Shaw & Walton, 1970). Later, the effect of stocking rate on pasture yield and composition may be of critical interest and only pasture measurements are then taken. However, since animal performance is almost invariably the central objective of a grazing experiment, animal production should be measured as directly as possible. Nevertheless, as measurements of animal production are not usually sufficient by themselves to achieve an understanding of the complexities of a grazing system, many attributes of pastures and of animals, as well as of animal products, should be measured.

Grazing may affect grassland in many ways. Plants are trampled and may be damaged to a degree which varies with species, stage of growth, soil type, type of animal, and soil moisture status (Campbell, 1966; Edmond 1966, 1970b). Dung and urine are distributed unevenly (e.g. Petersen, Lucas & Woodhouse, 1956; Hilder, 1964; Marsh & Campling, 1970). Grazing is uneven (Morris, 1969), leaving swards more variable than those which have been defoliated by machines. The respective quantities of different species consumed will change a pasture, because stock graze selectively (e.g. Weir & Torell, 1959; Lesperance *et al.*, 1960; Arnold *et al.*, 1966). The botanical composition may become quite different under grazing from that under cutting, even though nutrient status is kept as comparable as possible (Brockman, Shaw & Walton, 1970). The effects of grazing and grazing systems have been reviewed by Myers (1972). Both he and Humphreys (1966) emphasized that cutting experiments provide, at best, an imperfect guide to grazing practice.

These facts, and others outlined by Reed (1972) in relation to legume/grass ratios, have led some authors to conclude that experiments which evaluate grasslands by estimating ' yield ' under cutting are of little value for evaluating grasslands for animal production. Undoubtedly this extreme view has some validity, but it is not always progressive.

Few pasture plants or management practices have been subjected to rigorous tests by animal production, yet very substantial gains in production per animal and per unit area have been achieved (Morley, 1962). In general, these achievements have probably been based largely on cutting tests, usually not even under grazing, which gave useful guidance to fertilizer utilization, selection of plant species and potential yields, and carrying capacities. They were often coupled with some shrewd observations by farmers and others.

Pasture yields estimated during the course of a test of a grazing system can be said to constitute ' evaluation by cutting '. The quantity of pasture which can be harvested on the plots of such tests would be expected to be correlated with animal performance, because it reflects some function of the amount of feed which is surplus to animal requirements, or the animal's ability to harvest it. Therefore, it is not surprising that Scott (1968) and Joblin *et al.* (1972) found a ' general satisfactory ' relation between pasture dry matter and animal performance. But such a relation cannot be taken as real evidence that treatment effects can simply be evaluated from measurements of pastures. The relationship between the amount of pasture present and animal performance would probably vary between species and seasons of the year. Moreover, the pasture available may well reflect palatability rather than pasture growth, so that at times the more herbage there is present, the lower the performance of animals which graze it.

Because such bases of evaluation embody so many subjective elements and traps for the unwary, it is necessary at some stage to test new or modified material and practices in animal production systems. Given that tests at various stages are all part of a valid strategy for the improvement of animal production from grasslands, some combination of different types of measurements will maximize the benefits flowing from the resources available.

Intuitively, animal production tests will be the most important as productivity approaches the potential set by the resources of soil, climate, and animal efficiency. At the earliest stages substantial advances may be obtained by, for example, an effective strain of rhizobium, correction of a trace element deficiency, or the introduction of a plant that will persist and yield well. Although it would indeed be reckless to suggest that such advances belong only to the past, it would not be unduly rash to admit that they are likely to occur much less frequently in future. Progress now seems more likely to follow from fairly precise estimates of animal and plant responses to complete production systems. Therefore, the grazing animal must play an increasingly important role in the measurement of grassland productivity.

The simplest forms of grazing experiments are those that aim to compare variables such as varieties of pasture species, fertilizers or management practices in isolation. Eventually, the variables in question may also be evaluated in physical or conceptual animal production systems. As one moves through this sequence of experimentation, problems of time, resources, and personnel may become so great as the complexity increases that the number of treatments that can be investigated must become seriously restricted. An insistence on evaluaations in a complete animal production system may result in eventual failure to achieve the desired goal, because the experimenter is swamped by measurements, and possibly because of difficulties of interpretation.

Further progress must then be based on logic and understanding of the basis of soil-plant-animal systems, and the importance of the outputs of treatments for animal production systems. This logic and understanding must be soundly based on the *facts* of animal production, and must be continually verified by measurements of animal production on actual production systems.

Investigation of management practices, and the development of better management, must continually receive high priority in grassland research if farm profits are to remain reasonably high and agricultural products cheap. And as advances from mechanisation, land development and reorganization become less dramatic, research is likely to be increasingly concerned with improvements of the order of 10 or 20 percent in new systems of production. Such changes demand levels of experimental precision which would not have been appropriate in the past. The investigation of grassland productivity and management, therefore, demand designs in which the estimation of response surfaces will increasingly replace the testing of null hypotheses.

The dependent variable(s) of response surfaces must also become more complex as management becomes more sophisticated. It will no longer be sufficient, if in fact it ever has been, to analyse responses in terms of animal products per unit area. Other variables, such as conditions of the pasture, vulnerability to erosion, markets, labour demands and cash flows must impinge on biology. Not that the biologist must become a management expert, but at least he must be aware of the impact that the application of his results could have on management. To ignore this is to risk doing investigations that are irrelevant.

In summary, although animal production remains the keystone of grassland productivity, and it is thus necessary to include animal production in most serious attempts to measure grassland productivity, it cannot be regarded as a wholly sufficient measure, or one which is the most appropriate in all circumstances.

In animal production experiments, complex designs and detailed measurements can be prohibitively vulnerable to deaths of animals and other management hazards. A small experiment may be more useful than a large and complex one, because in large grazing experiments it may not be possible to measure the growth of pastures, the consumption of pasture by animals, or even small but perhaps important changes in botanical composition, accurately enough to enable interpretation to be unequivocal. A compromise must therefore be found between two extremes. An experiment which, because many measurements are involved, is so restricted that extrapolation beyond the actual animals and relevant environments is hazardous, may be of less value than a large experiment, with relatively few observations, covering a wider range of environments.

IV. DESIGN OF PASTURE-ANIMAL TESTS AND EXPERIMENTS

1. General

The design of tests and experiments which attempt the measurement of animal production from grassland must be based on the purpose of such experiments, (Willoughby, 1975; Walshe, 1975).

Firstly it is necessary to define the relevant production system; this may be an actual or potential system. In either case the definition should be no more rigid than is necessary for the design and the interpretation of results because, as the studies progress, new and previously unsuspected opportunities or limitations may come to light. Continuation of long-term experiments along an original plan may thus come to mean the continuation of the collection of information which is, at best, of limited relevance. New techniques of fodder conservation, new fertilizers, and even new breeds of animal or new cultivars of pasture, could render present tests redundant. The decision to change or leave unchanged the design during the course of an experiment may demand all the understanding, experience, and perhaps courage the experimenter can muster.

It is possible to provide for later modification in an experimental design. For example, if four replicates are used initially, an additional treatment can be imposed on two of the replicates, which still leaves a balanced design. If only three replicates were used initially, however, the new design would be more complex. Changes also demand an examination of possible outcomes to decide whether such decisions will continue to permit valid interpretation of results. Secondly, it is necessary to define the limits imposed by the design and the system to which the results are expected to apply. This definition, which may have to be modified after the experiment, can involve additional variables such as the geographical unit on a regional scale as well as the site and paddocks or portions of paddocks. Breed of animal, and animal management decisions (reproduction, marketing, classes of stock) should also be considered in the design, especially when prior knowledge indicates that these could be important.

Experiments in animal production from grasslands are usually performed to provide a basis for decisions. Inevitably, conflicts arise between the objective of maximising the precision of comparisons within the framework of a given experiment involving certain resources, and the value of the results of such an experiment in management decisions. For example, it may be much more valuable to establish a difference with $P > 0 \cdot 10$ over several sites and which is applicable to a wide region, than to have a statistically more precise experiment, the results of which are of little value for generalization. That is, at the design stage the kinds of decisions that might be influenced by the result of the experiment should be considered. Then the most sensitive comparisons should be associated with those decisions where errors could have the most serious consequences.

Factual knowledge need not be based only on results of formal experiments, and in the early stages of investigation collection of data from farms through surveys may be very useful. For example, the production of 500 kg beef per ha from a certain pasture may provide useful information if it were generally known or believed that 300 kg constituted a theoretical or practical upper limit. Such a result would immediately pose questions about the type of site, rainfall and many other variables which could have generated it, and the range of conditions to which it might apply. The development of further understanding might well require formal experiments.

The experimental unit of a production system in the measurement of animal production from grasslands must consist of an area of land and the animals grazing on it, as well as any auxilliary facilities for such management activities as supplementary feeding. This concept has been elaborated by Morley and Spedding (1968). Two approaches are possible. First the replicates of an experiment may be random selections of possible sites of the region of extrapolation. The variance of a treatment mean will then contain two components, one being the interaction between sites and treatments, the other derived from the variation among animals in each treatment. If (ST) is the interaction and A the variance among animals in a plot, the variance of treatment means = (ST)/s + A/(a.s.), where s is the number of sites and a the number of animals per experimental unit (plot). Usually the size of (ST) will be unknown. It could be large in most grassland studies which are designed to be applied to a number of farms in a reasonably extensive region. Since not many experiments are feasible per million cattle or per ten million sheep in any country, designs should usually aim to give fairly wide generalization rather than very precise answers applicable only to very special situations.

The most efficient use of total resources depends on the relative costs of groups of animals and plots of different sizes in the establishing and running of the experiment. Optimum designs can be formulated if one knows the relative costs and the relative sizes of variance components. For a given number of animals, variance is minimised by having one animal per plot. This may, however, introduce complications due to abnormal behaviour. Donnelly, Axelsen and Morley (1970) found no effect of flock size on the performance of weaner sheep down to three animals per plot, but Southcott, Roe and Turner (1962) reported that wethers in flocks of two produced less than those of flocks of four or more. In practice, since one seldom knows either costs or variances precisely, the usual compromise is an attempt to avoid problems of animal behaviour, to provide a reasonable sample of the environment, and to facilitate management.

An alternative approach, which may have much to recommend it, is to use partial replication and to confound replication with site. Thus, if one wishes to extrapolate to a region characterised by four main classes of site, one would locate at least one replicate in each type of site, with additional replicates on sites which represent areas offering major opportunities for using the techniques in question. The analysis may be found on the assumption that the sites are not random, and the differences between results from different sites can be used to sharpen understanding of the total system and its interactions, and to estimate the value of the treatments being studied for the total region of extrapolation.

Replications in a formal sense may be unnecessary if the design permits the use of regression analysis. Stocking rates, fertilizer levels, and many other management variables (including perhaps even botanical composition) can be varied systematically. Responses can be estimated by the usual regression techniques. Replication may be included or omitted, depending on the kind of precision required. This approach may be especially useful where information on responses and optimum levels is meagre. If one has little idea of what an optimum stocking rate or level of fertilizer application may be, evaluation of a treatment in an unreplicated experiment involving a wide range of levels may be the most informative. If one already has fairly precise estimates of optimum levels, but needs increased precision in the evaluation of some managerial treatment, replication of the treatment at stocking rates or fertilizer levels in the region of the optimum may be desirable.

If one knew precisely the shape of responses to variables such as stocking rate, one would not need to vary them. This is essentially the viewpoint of Conniffe, Browne and Walshe (1970, 1972) who emphasised that a fair assessment of treatments requires that they be compared over a set of stocking rates so that an optimum rate can be estimated for each treatment. They suggested that the optimum could be predicted from an observed production at a single stocking rate *provided* " there was a known relationship between animal production and stocking rate ", and " that the same equation will apply on any experimental site and to any set of treatments ".

This is an extension of approaches proposed by Mott (1960) and by Owen and Ridgman (1968), and is inherent in some other designs. It is obviously a correct approach, provided the explicit and implicit assumptions are valid. Unfortunately, one is seldom likely to know when the assumptions are valid, or to have a precise estimate of response to stocking rate, or the shape of responses to the fertilizer level, subdivision, fodder reserves, supplementary feeding, or other variables which may deserve as much attention as stocking rate in this context. It would be inappropriate here to discuss details of designs such as randomized blocks and factorials, as these are adequately considered in the statistical literature. Split-plot and change-over designs present some special features which are worth mentioning briefly.

2. Split-plot designs

Split-plot designs are well known in field crop experiments, but less so in animal production. Almost invariably one would wish to know the effect of a grassland variable on different types of animal (age, sex, breed, or species) rather than on one special type. As a general principle, the more intensely sites or animals are selected from a single population, the less justified will be extrapolation to a substantial proportion of an industry. If the animals in an experimental unit are of different breeds, or vary in some other way, little is likely to be lost and useful information may be gained. Breed × treatment interactions may be found to be quite small or quite large. If the former, it is good to know. If the latter, it may be of critical importance to know. For example, Donnelly *et al.* (1970) examined the effect of flock sizes of 3, 9, 27 and 135, and of pasture management with experimental units composed of three breed types. No significant difference between flock sizes was found, and interactions of breed with other variables were slight and not significant. Since the experiment was performed with mixed breeds the results are of far more value than if it had been done with a single breed. Had the performance of one, but not all, breeds been affected by treatment, the experiment would have been of special interest for obvious reasons. In this case the use of the split-plot design for breeds cost little and was of great benefit.

3. Change-over designs

Where the individual animal can be the experimental unit, for example in evaluating different pastures for milk production with dairy cows, the effects of differences between animals on the variance of differences between treatment means may, in some circumstances, be reduced by ' change-over ' designs. (See Lucas, 1960; Patterson & Lucas, 1962; Davis & Hall, 1969; Stobbs & Sandland, 1972). Their efficiency depends primarily on : (a) high permanency of differences between animals over the total duration of tests, (b) small animal × treatment interaction, and (c) small carry-over effects of treatment from period to period. If permanent differences between individuals are large, such designs may be especially useful.

Unfortunately, carry-over effects from one treatment to another seem likely to limit the validity of change-over designs, and indeed the interpretation of animal responses in general, especially where changes in liveweight are an important consideration. Changes in body composition during periods of liveweight loss, and of weight regains during realimentation, may grossly distort estimates based on liveweight, of energy storage or loss (Donnelly & Freer, 1974; Drew & Reid, 1975 a, b). Although it may be desirable to measure energy storage, this will often be impractical in grazing experiments. If large variations in nutritional status are imposed, some allowances for changes in body composition may be necessary for interpretation of animal performance. Perhaps a more complete understanding of the processes involved will permit interpretation without extensive measurements.

Grazing experiments are beset with many physical limitations and are unlikely to give unequivocal results over more than a minute sample of the possible systems which might incorporate some new technology. Therefore, formal experimental comparisons using conventional designs and replicated experiments will often be generally inadequate in the measurement of animal production from grassland systems. Recognition of this inadequacy has generated alternative approaches involving model building and systems analysis which will be discussed later in this chapter. The need for accurate information for such activities, and the cost and size of most experiments with grazing animals, emphasize the importance of designs which are efficient and sufficient, and which are not vulnerable to the minor disasters which often beset experiments in the field.

V. CHOICE OF PASTURE SPECIES FOR ANIMAL PRODUCTION

1. The criteria for acceptance of material for testing

The choice of pasture species may depend primarily on animal production, but many plant attributes can be important if not critical. Since animal production tests may be lengthy and costly, it should be axiomatic that species or cultivars accepted for animal tests should have already given clear indications that they are highly likely to be useful. Criteria for acceptance might include :

(i) Freedom from deleterious physical structures or harmful chemical compounds (toxins, oestrogens, flavouring agents).

(ii) Ecological adaptation to the grazing environment, including other plant species in the pasture community, pests and diseases.

(iii) Agricultural adaptation, including seed production or vegetative propagation, economical establishment, and the ability to withstand grazing in some feasible system.

(iv) Some prospect of offering advantages over lines in current use, based on yields in swards, preferably under grazing, and information on digestibility and acceptability.

2. Objectives of testing

A line for testing may be selected for some special purpose, or to replace or complement a cultivar or cultivars already in use. The objectives should determine the techniques to be used in testing. Thus if *Medicago sativa* in dryland pasture is considered to be especially useful for growing and fattening young animals but of dubious value for older breeding stock, comparisons among lines for testing should be on the basis of performance in growing and fattening livestock. This may, of course, not preclude the testing of lines which could be especially valuable for breeding stock. Tests should be performed in such a way that results are interpretable in terms of the total production system; hence the demands of the production system and not the feelings of the experimenter should determine when animals graze the species under test, when they are to be replaced, and what measurements are to be made on the animals and on the pastures (Axelsen & Morley, 1968).

3. Management of grasslands under test

Grasslands usually consist of rather complex communities of plants which may, however, become greatly simplified under grazing. A pasture species for testing may be used as part of a moderately complex community of different grasses and legumes and other species including weeds. Alternatively, it may be used as a pure stand, especially if high levels of nitrogenous fertilizers are applied. It is important to distinguish clearly between tests of pure stands and tests of pasture mixtures which include the line being studied. Each may yield relevant results, but relevant to different situations. A particular routine of grazing management of the pasture may be optimal for one line but far from optimal for a second, which may therefore be penalised in the comparison. For example, different lines of *Medicago sativa* or *Leucaena leucocephala* may all require some form of rotational grazing if they are to survive, or to be productive, for more than a year or two. But because different lines develop at different rates, the cycle length of an optimum rotational system may vary greatly between lines. Similarly, lines may react differently to stocking rate, cutting for hay or silage, extremes of moisture, weed competition, trampling by stock and many other components of grazing systems. The accumulation of information on such reactions is part of the test of a line, and some lines may be discarded without further tests.

Many of these measurements cannot be made without the use of grazing animals, but they are essential to the understanding of the potential role of the lines to be tested. Measurements which do not involve animals, measurements of responses to grazing, and measurements of animal production may all contribute useful information. A compromise may therefore be needed to achieve the best use of available resources. There does not seem to be any simple criterion for allocation of priorities to achieve the best compromise. Some understanding of the likely reaction of lines to management could be required before lines are included in animal production tests. This may have been obtained from experiments designed to estimate such reactions, or it may be based on reasonable assumptions founded on knowledge of similar material.

Alternatively, resistance to treading or reaction to weed competition for example, could be studied in small, intensively measured plots. For such studies to be relevant, they should be done in such a way that treatments may be regarded as valid indications of what would be likely to happen under actual grazing systems. It might be almost irrelevant, for example, to discover that one cultivar withstands prolonged defoliation to ground level better than another if such treatment is not likely to be imposed on a significant fraction of grasslands in real animal production systems.

If important differences between cultivars in management requirements are known to exist, management should at least approach the optimum for each cultivar. This may, of course, complicate tests sufficiently to make them impractical, but neglect of this requirement may lead to irrelevant, or indeed misleading, data. Fortunately, it is unlikely that such interactions will be so important that they complicate many tests. Moreover, the management of any tests involves departures from what is likely in commercial practice. Therefore, if a line to be tested is highly sensitive to management and performs badly as a consequence, the verdict of the test may not be unreasonable. Resistance to mismanagement is an important requirement of grasslands.

4. Extent and duration of tests

Interactions between genotype and climatic or edaphic environment are undoubtedly widespread in grassland species. If such interactions are large relative to ' main effects ' at the farm or paddock level, the problems of selecting species by tests on animal production seem almost insuperable. Fortunately, interactions although real, seem generally unlikely to be as large as main effects, and may well be sorted into reasonably homogeneous sets related, for example, to wet and dry years or to fertile and infertile sites, which may then permit the selection of species for kinds of environments rather than for specific situations. In commercial practice, adaptation to a particular environment in a grassland is almost an unreal concept because of the ever-changing nature of the environment, except perhaps in highly controlled situations such as irrigated pastures maintained with high inputs of fertilizer. In the real world, pasture genotypes must be sufficiently adapted to a fairly wide range of conditions to be able to persist in the face of a range of climatic, edaphic, management and biotic challenges.

The optimum extent and duration of tests will depend on the relative importance of main effects and interactions, and the costs of duplication in space and time. This statement cannot, however, be very helpful because the main effects and interactions cannot be known for any region without doing work on an adequate sample of that region; in addition, none of the costs is likely to be known precisely. Because of these uncertainties the evaluation of genotypes may best be undertaken by the collection of data involving important variables which influence animal production, in the hope that a better understanding of these variables might permit extrapolation to a region considerably larger than that which might feasibly be covered by tests of actual production systems.

Applying this strategy, estimates of animal production should, where possible, be supplemented by data on digestibility of material eaten, diet composition and availability at intervals, perhaps supplemented by data on mineral and protein content and palatability. Meteorological and soil data may be especially important. This aggregate of measurements should enable a fairly complete picture of the main determinants of animal production to be drawn. Estimates should be obtained at stocking rates which seem likely to bracket an optimum stocking rate. Thus, differences in availability are introduced, which may aid the extrapolation to sites of different productivity and to years of different potential growth.

It seems unlikely that tests at a large number of centres would contribute much more than tests at a few carefully chosen centres over sufficient years to enable sampling of a range of yearly weather and covering most of the variation to be expected over a wide region. If edaphic specialization is important, tests on a number of sites at one centre could be more revealing than tests at many centres, and would probably be much cheaper and simpler. The significance of both quantity and quality in determining animal production needs to be appreciated if the relative values of different genotypes are to be understood. Tests that do not evaluate both quantity and quality may be of little value, and they are in danger of being irrelevant unless performed within the framework of a real production system.

5. Release of tested material

Tests by animal production have not been used to characterize more than very few of our pasture species. Animal production tests are unlikely to be sufficiently precise, nor will the differences be large enough, to justify more than an occasional asterisk opposite the mean square for ' varieties ', or its equivalent. Differences as high as 10 percent are unlikely to be statistically significant, yet they could be of great economic importance. This being so, material need not really be outstanding, in terms of animal production, before being released for general use.

The optimum balance between a conservative strategy which insists on extensive evaluation by tests on animal production, and more adventurous strategies which do not require more than the demonstration of satisfactory animal performance, probably depends largely on the stage of pasture development in the region. In regions with highly developed technologies, tests should probably be far more stringent than in those where pasture improvement is still in course of development.

VI. STOCKING RATES

Modification of management by the adoption of new techniques frequently changes the amount and quality of pasture grown, and hence the feed available to grazing animals. The use of fertilizers, new pasture species, fodder conservation, and subdivision are familiar examples. As a consequence of these changes the number of animals that can be maintained on a given area at existing levels of production may be greatly increased. In addition, the production per animal at a given stocking rate may be appreciably altered. An understanding of such interactions led McMeekan (1960) to emphasise the importance of stocking rate as the main determinant of grassland production. The adoption of a new technique, therefore, calls for a decision on whether to change the stocking rate. In some circumstances the new technique may be worthwhile only if the stocking rate is changed. The decision must depend not only on the effect of the technique on production, but on the objectives of the decision maker who is the one to define what he means by ' optimum '.

Numerous experiments to estimate the optimum stocking rate have been proposed and some carried out, but no design seems to be completely satisfactory. The basic problem is to estimate a response surface in a system subject to variation in space and time, sampling errors, carry-over effects and curvilinearity. In addition, rules must be formulated for management decisions (e.g. when to give supplementary feed) which are reasonably in accord not only with what is feasible, but near optimal, in the context of present or future practice. Furthermore, the results obtained will need to be translated into decisions which can be applied to a range of sites and years. If the definition of production is extended to include profitability, conservation of resources, and utility (see McArthur & Dillon, 1971) the question becomes far from simple. The same biological information is needed for a wide range of objectives and for most sensible definitions of ' optimum '.

Estimates of responses to stocking rate are required for at least the following :

(a) Production per head
  (i) Physical — wool, meat, milk, etc.
  — natural increase
  (ii) Financial — value of products
(b) Cost per head including deaths, supplementary feed and additional labour.
(c) Stability of pasture—soil losses, residues at different times.

If one knew the precise shape of response curves to stocking rate it would be sufficient simply to determine one's location on this curve by means of two levels of stocking or perhaps only one. Unfortunately, although general shapes of response curves have been postulated by Mott (1960), Riewe (1961) and Jones and Sandland (1974), a general response curve has not yet been sufficiently validated to justify such a method of estimation. Curvilinear responses are predicted because performance per animal is likely to be near maximum at all stocking rates up to that level at which feed quantity starts to become limiting. This would occur at different stocking rates at different times of the year, and would also vary between years. When responses are summed over a whole year, or several years, they are most unlikely to resemble the discontinuous functions described by Mott (1960), and depicted in several other publications.

Some biologists may object to the inclusion of economic terms in the list of responses to be estimated. Since animal production from grasslands is almost invariably aimed at financial rewards, it seems unrealistic not to include these among the criteria for decisions relating to stocking rates, or indeed to any other production functions. Experiments should at least be designed so that economic parameters may be included in the analyses of results.

The response of production per head to stocking rate are likely to be curvilinear and, even if production per head and costs per head were linear with stocking rate, production per ha and profit per ha would be curvilinear. This is evident from the following table as well as from simple algebra.

	Stocking rate				
	1	2	3	4	5
Production/head	10	9	8	7	6
Production/area	10	18	24	28	30
Costs/head	3·0	3·5	4·0	4·5	5·0
Costs/area	3·0	7·0	12·0	18·0	25·0
Profit/area	7	11	12	10	5

Because curvilinearity of responses is to be expected, many experiments designed to measure the effect of some treatment include three or more stocking rates. Unfortunately, most realistic grazing experiments cannot be large enough for curvilinear effects to be estimated with useful precision. Even the estimation of linear regressions with sufficient accuracy to determine optima within 10 percent of a true value is likely to be beyond the resources available.

General approaches to estimation of optimum stocking rates and the production functions from available data have been described by Heady and Dillon (1961) and elaborated by Chisholm (1965), Byrne (1968), Firth, Evans and Bryan (1975) and others. The calculation of gross margins seems to be appropriate for this purpose, but techniques such as inventory analysis may have special value if the financial stability of the enterprise is to be examined. Published estimates of responses to stocking rates, including those using utility functions (McArthur, 1970; McArthur & Dillon, 1971), from studies in which fixed rates of stocking have been applied throughout the grazing year permit some generalizations.

Over the broad range of stocking rates which are likely to be commercially significant, production per head is close to a linear function of animals per unit area (e.g. Hart, 1972; Jones, 1974; Jones & Sandland, 1974; Sandland & Jones, 1975; Jardine, O'Brien & Frew, 1975). Estimates of maximum gross margins per unit area have generally assumed a linear relationship between animal production and stocking rate (e.g. Chisolm, 1965; Byrne, 1968; Blackburn, Frew & Mullaney, 1973; Jardine, 1975). Yet, considered over a wide range, Mott's (1960) concept of a curvilinear response of stocking rate on animal liveweight gains has considerable merit and finds some support in the results of Bennett *et al.* (1970).

The critical question is therefore whether departures from linearity lead to serious biases in the estimation of economically optimum stocking rates. This seems unlikely because the economic optimum must lie above any 'critical value' below which animal performance is not affected by stocking rate, and will generally be below levels at which inputs, such as emergency feeding, are likely to rise exponentially or production per head fall exponentially. The responses will therefore be approximately linear in the region of the greatest interest.

These considerations are important in the design of experiments. The experimental estimation of an optimum stocking rate may be most efficient if only two levels of stocking are compared. The lower of these might be some 10 to 20 percent below a presumed optimum, the higher some 10 to 20 percent above that level. This should permit the maximum efficiency of estimation of a linear response about the optimum. The inclusion of treatments at the presumed optimum would contribute less to the accuracy of the linear component than would treatments above and below it. The inclusion of more than two levels would be unlikely to provide an estimate of the quadratic component of sufficient accuracy to justify the resources used. It is important to recognise in this context that the term optimum as used here is the economic optimum, not the point of maximum production as used by Jones and Sandland (1974) and others.

Objections may well be raised to this approach on the grounds that one would not know when the lower stocking rate was on a flat portion of the response of production per head to stocking rate. Superficially, this agreement seems unanswerable. However, information accumulated in the last decade or two should provide a fair indication of where an optimum is likely to lie for most temperate pastures and for some types of tropical pastures (see 't Mannetje, Jones & Stobbs, 1976).

As a first approximation, the optimum stocking rate on improved mesophytic pastures is likely to be approximately those in the following table :

Likely Optimum Stocking Rates (Animals/ha)

Rainfall (mm)	*Merino Ewe*	*Romney Ewe*	*Beef Steer*	*Beef Cow*	*Dairy Cow*
400	5·0	?	0·7	0·3	?
600	10·0	6·5	1·4	1·2	0·5
800	14·0	10·0	2·0	1·7	1·0
1000	17·0	12·0	2·4	2·0	1·5

If rainfall is badly distributed, the area may be unsuitable for dairy cows or lamb production. In addition, the above stocking rates should be discounted, perhaps by a factor which is approximately the square root of the proportion of the year which provides conditions unsuitable for plant growth. Thus an area with a rainfall of 800 mm, but a growing season of 6 months might have an optimum of about $0 \cdot 71 \times 14 \cdot 0 = 9 \cdot 9$ merino ewes/ha.

The interpretation of the biological information and its use in decision making on the farm presents some major problems. Analyses such as those referred to assume variable costs to be constant, or to be continuous functions. This is valid only where enterprises are very large. On the majority of farms the costs of labour may be increased or decreased, depending on whether an additional unit has to be added to the work force, or whether surplus labour can be used to care for additional livestock. The costs of many facilities will also be ' lumpy '. The optimum will therefore vary from farm to farm even where climate, soils, animals, management and pastures are similar. This in no way invalidates measurement of the responses of components of animal production to changes in stocking rates. It does, however, imply that the range of interest is rather more broad than might be suggested by continuous models. For example, on a small farm where a farmer could care for twice the present number of livestock, the optimum could be substantially higher than 25 percent above the optimum on a large farm where labour is fully used.

The biological information required for decision making on the farm seems to be largely independent of farm size, although the range of interest will be greater for smaller farms. The decision process must however take into account the variable-cost inputs, and these will be almost specific for a given farm. The extent of the range of interest should be considered when designing experiments. Practices such as fodder conservation may, depending on the cost, increase the optimum stocking rate to some extent. Sites with poor soils, steep hillsides, or poor drainage, must obviously have lower carrying capacities. I cannot emphasize too strongly that the figures above should be regarded as no more than aids to good sense and background information.

Estimation of the optimum stocking rate on grasslands which, because of low rainfall or soils with low water holding capacity or low fertility, have very low carrying capacities, presents special problems. The optimum stocking rate could be that at which the balance of edible plants and those contributing substantially to the stability of the grassland can be maintained in a stable community. This must involve some form of management because, undoubtedly, some of the naturally occurring species will not withstand grazing (e.g. Ebersohn, 1966; Moore & Biddiscombe, 1964; Acocks, 1966; Roberts, 1967; Moore, 1970; Smoliak, 1974).

The replacement of species which cannot withstand grazing by resistant species may not lower the animal production potential of the grassland or significantly reduce stability. Sometimes it undoubtedly does, but the assumption that a climax vegetation is to be preferred for animal production cannot be accepted as axiomatic; nor need the replacement of perennial species by annuals necessarily lead to a less stable community or lower animal production (Heady, 1961).

The evaluation of residues has been attempted by Hooper and Heady (1970) for the annual range of California. They attempted to value ' mulch ' in terms of its equivalent in feeding value as herbage, and its value in preventing erosion and by its effect on plant growth. There are obvious gaps in understanding here because of different susceptibilities of different soils to erosion, variations in the costs of damage by erosion, and uncertainties about the effect of residues on pasture production.

The value of mulch in increasing herbage production will vary greatly from year to year, so that the use of a single substitution value does not seem appropriate in this context, since at times the mulch will have zero value because there will be a surplus of forage. A more logical evaluation would be to penalize too high a stocking rate by the cost of removing animals from the pasture to a feed-lot and feeding them at maintenance levels. This would be necessary more frequently at high stocking rates. The analysis could be taken much further to include fattening stock for sale as a part of a feed-lot production system. As the feed-lot becomes a larger part of the total system, the evaluation of the range would be oriented increasingly through some kind of ' put and take ' system of management, with rules on amounts of residues to be left on sites with different characteristics.

The precise estimation of the effect of stocking rate or other management variables on binomially distributed attributes such as natural increase or mortality is unlikely to be feasible in experiments. The achievement of 95 percent confidence intervals of about 10 percent would require about 250 animals per treatment. This has two consequences. Firstly, if binomially distributed attributes are to be estimated this must generally be done indirectly through correlated variables and models of relationships, as the use of weight at joining to predict fertility as by Morley, Axelsen and Cunningham (1976). Secondly, if the experiment is performed with small groups, sampling errors will mean that the balance between treatments could be distorted. The effective stocking rate of 8 cows and 8 calves on one plot may be quite different from that of 8 cows and 4 calves on another plot of the same treatment.

This problem may be overcome in part by switching animals between plots belonging to the same treatment, provided it does not unduly bias plot effects. For example, the two plots each with 8 cows, but one with 8 calves and the other with 4, could be reorganized to give 8 cows and 6 calves on each. Where differences from sampling occur between treatments, difficulties are more serious as one cannot switch animals between treatments even if one is almost sure that differences are due to sampling. There is no obvious escape from such dilemmas.

The choice between ‘ animals per unit area ’ or its reciprocal as an expression of stocking rate is discussed from time to time in the literature. Area per animal might be preferred if pasture growth were not affected by stocking rate. The amount of food available per animal would then be a simple linear function of stocking rates. However, evidence from Campbell (1966 a, b, c, d), Edmond (1966, 1970) and Vickery (1972), as well as theoretical considerations revolving around leaf area indices, suggest that stocking rate might affect pasture growth.

As stated by Shaw (1970), comparisons among stocking rates need not be based on any mathematical series. For demonstration purposes, there may be some value in including stocking rates which are in general use. Otherwise, the criterion for concentration of stocking rates over any part of the range might be the value of the information to be obtained. This is likely to be slight at stocking rates that are very high or very low, and maximum at commercially optimum levels, not necessarily at present commercial levels.

The response of production per head to stocking rate will vary between sites and between years (Cannon, 1972). Therefore, the results of a given experiment must be limited because they are specific to the particular site(s) and year(s). Attempts to overcome this limitation are increasingly based on models of production systems which take into account such variables as weather, soil structure and pasture type. Since the construction of such models is an important activity at many institutions, pasture management research workers may need to be increasingly aware of this rapidly developing field.

VII. RESPONSES BY ANIMAL PRODUCTION TO FERTILIZERS

Pastures on most soils respond to nitrogenous fertilizers, on many to phosphorus and on some to potassium and sulphur, depending on soil fertility, rainfall, use, etc. Although nitrogen deficiency may be largely corrected by legumes, these will usually grow well only if their nutrient supply is augmented by fertilizers. In the early stages of the development of an improved pasture, superphosphate and trace elements plus a legume usually induce dramatic increases in plant growth. As fertilizer application continues, the rewards from phosphate tend to decline. Eventually, a stage is approached at which only small applications of longer acting fertilizers are needed to maintain nearly the maximum growth possible in the environment. By this time, however, other deficiencies may have become more evident, and responses to quick-acting fertilizers such as N become increasingly significant. The response of pastures, and hence animal production, to fertilizer inputs thus involves questions of immediate response and carry-over effects.

Barrow (1967) reviewed the effects of grazing on the nutrition of pastures and discussed the significance of the mobility of ions in the soil and the consequences of fixation and leaching of nutrients. Nitrogen acts primarily in the short term, phosphate in the long term, whereas potassium and sulphur are intermediate in this respect. The problems of animal responses to a mobile element will now be considered, with phosphorus used as an example of a relatively immobile element, and some problems involved in estimating optimum levels of application will be discussed.

The efficient use of fertilizers is an important economic goal. To apply too much may waste money; to apply too little may reduce production. The problem of assessing the economic requirements for fertilizer becomes more acute as experimentation proceeds from crop agronomy on selected uniform soils to grazing trials on variable sites. In the former, cultivation, herbicides, pure stands, and standardized sowing and harvesting procedures are aids in preserving uniformity in responses to applied fertilizer. However, the pasture agronomist has to contend with responses by animal production to responses by pastures of different botanical compositions on variable sites and during different seasonal conditions. Long-term responses must take account of changes in botanical composition, which may not be quickly reversible, as well as responses in plant growth.

1. Evaluation of nitrogenous fertilizers

The application of nitrogenous fertilizers to developed pastures introduces the problem of evaluation of short-term responses in growth and botanical composition by animal production. This is the subject of a substantial proportion of research on pasture productivity in Europe, North America and some parts of the wet tropics.

Holmes (1968) reviewed the use of nitrogen for cattle pastures and Conniffe, Browne and Walshe (1970, 1972) discussed experimental designs which are relevant to dairy and beef cattle production systems and which, in principle, might apply to other forms of production. As emphasised by them and others, the use of nitrogen, if it is to be fully effective, may require a number of adjustments in management. Stocking rate, in particular, will require adjustment to the modified level of pasture growth (Browne & Walshe, 1968; Marsh & Murdock 1974). In addition, changes in botanical composition could mean changes in relationships between pasture growth and animal production (Reed, 1972; Bryan & Evans, 1973), fouling and treading may become more important, and new optimal levels of conservation and supplementary feeding could be generated. The estimation of multi-dimensional response surfaces which can take into account more than one or two of these management variables would simply not be feasible.

The effect of nitrogen will vary through the year following variations in soil moisture, temperature and radiation. The response is expected to be least when the need for additional growth is greatest. Hence, in evaluating nitrogen responses by animal production, strategies should be developed so that fertilizers are applied only at times when increased yields are likely to be sufficient to be economically justified. Henzell and Stirk (1963) and Jagtenberg (1970) have made some tentative moves in this direction, but a more thorough development is needed. Since most responses to nitrogenous fertilizers occur almost immediately but are of relatively brief duration, research on decision-making in systems of production could be relatively short-term if effects on botanical composition are reasonably well understood. An example may help to clarify this viewpoint.

In Eire, Browne (1966) measured liveweight gains of bullocks on a ley on which several levels of nitrogenous fertilizer were applied. Responses to the highest level were substantial (100 percent of control) at the beginning of the experiment, but they declined steadily to 21 percent within five years. This decline in response was not due to decreased production at high levels of fertilizer application, but to a steady increase in liveweight gains per ha in the cattle on the control treatment and in those on pastures receiving the lowest levels of N. The productivity of the ley at maximum levels of fertilizer was relatively constant. Another experiment on an old pasture gave results similar to those on the ley, except that the control seemed to asymptote to higher levels of production more quickly (Browne, 1967). Since these were ‘ put and take ’ experiments, an economic evaluation does not seem feasible without much more information on the total system. Furthermore, the ‘ put and take ’ experiments gave liveweight gains per acre which would probably be unattainable in practice, and responses to N which were very much greater than those achieved in a fixed stocking rate experiment. The information needed to explore alternative strategies would include studies on conservation of varying portions of the grassland system, the effects of fertilizer on the quality and the quantity (area!) conserved, the use of conserved material, and the responses to fertilizer applied at different times.

An experiment in a sharply contrasting environment illustrates the influence of nitrogen on animal production through the botanical composition of the pasture (Davies, Greenwood & Watson, 1966; Greenwood, Davies & Watson, 1967; Davies & Greenwood, 1972). Nitrogen at three levels was applied over four years to a newly sown pasture of subterranean clover and annual grasses on a sandy soil in the mediterranean environment of south-western Australia. Two stocking rates were used. Some response in animal production to the fertilizer was obtained in the first year. Thereafter the response was slight, or even negative. The authors attributed this result to the changes in botanical composition which were caused by the fertilizer. Their evidence suggests strongly that the nutritive value of the pasture depended heavily on the clover available rather than on the dry matter present. This is in accord with evidence such as that summarized by Reed (1972). A further contrast is reported by Tadmor, Eyal and Benjamin (1974) from semi-arid pastures in Israel. These pastures respond markedly to N fertilizers; also, the N fertilizers are not leached out of the soil but remain available to plants in subsequent years. Although N fertilizer is not economical in drought years, the fertilizer is not wasted and becomes economical in later years provided stocking rates are sufficiently high that surplus growth can be utilized. That is, N cycling may be critically important (Till & May, 1973).

2. Phosphatic fertilizers and animal production

The response to phosphorus is much more prolonged than that to nitrogen, and residual effects may persist for years, depending on the soil, rainfall, the form of production (e.g. milk or wool), and possibly on the stocking rate and pasture species present. The actual response at a given time will depend on the amount available to the plants, which in turn will depend on the amounts originally available and the sum of the amounts remaining from previous applications.

The effects of grazing on the nutritive requirements of the pasture, the cycling of nutrients in the soil/plant/animal system and the concentration by animal transfer of nutrients to portions of the total area mean that the grazing animal is a necessary part of the system to be examined. Since the objective of fertilizer application virtually always involves money, the value of animal production as well as that of land improvement must be estimated, or at least data provided to permit their estimation. However, since the effects of each phosphate application may extend over years, experiments must be continued until treatments can be evaluated as steady-state systems.

The measurement of responses in terms of animal production involves two kinds of problems. The first, which may be by far the more important in relatively underdeveloped grasslands, concerns the strategies for development of improved pastures, or in older agricultural areas from leys or from worn-out crop lands. The second concerns the estimation of the maintenance requirements of established pastures for fertilizers.

(a) *Development of improved pastures*

The development of improved pastures is a problem concerning the use of limited resources to generate income and to increase those resources. An example may clarify this general statement. If improved pastures were to be established on 100 ha of worn-out crop land, the resources available could be spent initially in rapid and intensive improvement of a small portion, or less intensively, on a slower development of the whole area. In the context of fertilizer use, a small area of a ley, such as that studied by Browne (1966), could be heavily fertilized and stocked, or the total area could be given a basic fertilizer and developed slowly over five years at a lower average stocking rate. An example is discussed by Morley (1968a).

The design of experiments to synthesise an optimum strategy of development should, especially in erratic climates, be based on recognition of the fact that the rate of development of improved pastures may depend heavily on seasonal conditions. It could, therefore, be desirable to commence one or more new replicates in each of several years. An adequate system might include two or more grazing intensities adjusted to keep up with the increases in productivity of the pastures, and three or four replicates commenced in different years.

Obviously such an experiment, if it is to be large enough to give clear guidelines, would seldom be feasible. A first approximation may, however, be obtained by estimation of the effects of fertilizers on the growth rates of pastures at various stages of development on relatively small, commonly grazed plots under different intensities of grazing. According to Brockman, Shaw and Walton (1970) this may be feasible for phosphorus and similar relatively immobile elements, but nutrient transfer could invalidate such tests for the more mobile elements. Inevitably, the emphasis shifts from examining many whole systems to estimating the most important variables for comparisons of models of systems.

(b) *Maintenance of fertility and improved pastures*

For the maintenance of improved pastures, the question that should be asked is " What animal management and fertilizer application system will maximize the objective function (profit, etc.)?" The question " What rate of application will maintain the phosphate status of this pasture?" is not adequate because, for example, different stocking rates, levels of conservation, and perhaps other variables would be at different optima with different levels of phosphate availability.

Again, the experimental exploration of all possible systems is scarcely feasible, even at only one site. Probably the largest attempt to estimate fertilizer requirements is that reported by Joblin *et al.* (1972) from New Zealand. Eight different factorial fertilizer × stocking rate experiments with one to six replications and 12 to 52 paddocks were described. A single experiment by Scott (1968) with two levels of fertilizer, two stocking rates and five replicates, stocked with breeding ewes, also indicated the resources needed to obtain an admittedly rough estimate of maintenance requirements. Both Scott (1968) and Joblin *et al.* (1972) found that, at least at higher stocking rates, there was a good correlation between pasture dry matter response and livestock performances.

The trials reported by Joblin *et al.* (1972) have contributed substantially to our understanding of the limitation of such experimental approaches. The relative merits of ewes and lambs, young sheep and dairy cows for measuring fertilizer responses depend on the system to which such results are to be applied and the vulnerability of results to management decisions. Joblin and his colleagues considered that management decisions should be minimized, but that the results from trials with breeding stock, which require continuous high levels of experimental management, could be highly vulnerable to management. The authors cast doubts on the value of a uniformity trial period prior to commencing the main experiment since the aims of the biometrician (presumably anxious to use covariance), and the desire of the experimenter to gain experience with the site and with the techniques to be used, may conflict if the experience results in changes in the experiment. Joblin *et al.* (1972) suggested that the main value of such trials is likely to lie in the ancillary information, such as effects of stocking rate on animal health and on the botanical composition of the pasture, and as preliminary information which might lead to a better understanding of site characteristics and pasture productivity. They also suggested that such trials have merit as sources of data from a range of environments, which may be useful in building models and developing simulation studies of animal production systems.

However, the collection of quantities of data from many and extensive field trials seems unlikely to be the most efficient approach to estimate the parameters needed. At least it would seem likely that the first stage in such an investigation should be the development of models and, using sensitivity and other tests, to allocate priorities to the estimation of different parameters, and then to perform the experimental work needed to estimate those which are most important.

3. The value of stocking rate × fertilizer experiments

Stocking rate × fertilizer experiments are unlikely to give generally useful estimates of optimum fertilizer-stocking rate combination or of the maintenance levels of fertilizer applications. The main reason for this is that the description of a response surface is unlikely to be precise, except possibly for one site over a limited set of years. Joblin *et al.* (1972) suggested that a recommendation was unlikely to be justifiably more precise than 'between 125 and 375 kg/ha of superphosphate'. Therefore, a particular recommendation, no matter how obtained, will be an underestimate or an overestimate.

In other words, the usual experiments tend to consider static situations; but the decision-making world of the farmer is dynamic. We can divide the fertilizer response of pastures into two major areas which pose two questions. First, what is the response in plant growth to the application of fertilizer? Second, what increase in animal production may be obtained from this plant response? The prediction of responses to fertilizer in different climates over a wide range of soil types having different fertilizer histories involves many problems. Soils vary widely over most pastures and differences in previous fertilizer applications, climatic variations, and grazing and cropping histories add further hazards. Prediction may be based on empirical tests. But the farmer may need tests of every paddock, even parts of paddocks, if he is to base decisions on such tests. Therefore, careful plot work and meticulous harvesting, so important to much agronomy, are out of the question at this level. Strips of fertilizer across paddocks or plots, especially if protected from grazing for some period, may give some indications of quantitative and qualitative plant responses to fertilizer which apply to the particular areas in question, and minimise assumptions. The responses may be measured by devices such as capacitance meters, cut quadrats and double sampling (see Chapter 4). Most people can observe by eye differences in yield of 10 to 20 percent between adjacent plots. If differences are evident, the next question follows.

If one can see or measure a response, does this mean that the application of fertilizer will give worthwhile responses in animal production ? If economically useful responses can be detected by eye they may be valuable in themselves or in supplementing other aids to predictions, such as soil tests, information on previous fertilizer application, rainfall, and botanical composition. One must also estimate the probabilities of whether, if a response is not observed, failure to apply fertilizer will adversely affect animal production.

The evaluation of a response by the pasture, through animal production, is a much more general question than the prediction of plant response. It is also more complex, as the value of the response must depend on management, the season, the cost of fertilizer, and the return from produce. Increased pasture availability through fertilizer may do no more than add fuel for grass fires. The application of fertilizer may also result in the herbage grown being more heavily grazed because it is more palatable (Ozanne & Howes, 1971 a, b). Botanical composition, if it is affected by fertilizer or stocking rate, as has been very evident in some experiments (e.g. Bryan & Evans, 1973; Wolfe & Lazenby, 1973) but not in others (Carter & Day, 1970), could affect animal performance through variations in the quality of the diet.

Increases in pasture growth brought about by the application of phosphate may both increase the carrying capacity at a given level of performance, and increase the animal performance at a given stocking rate. If the stocking rate is low, the extra returns may not pay for the costs (see Morley, 1974).

A first approximation to measuring the animal response might be through the results of comparisons of stocking rates, which essentially spread the feed produced by the pasture among different numbers of animals. This may permit inaccuracies by failing to take account of effects of stocking rate on pasture needs or growth response or changes in intake, or the value of feed, which follow applications of fertilizer (Ozanne & Howes, 1971 a, b; Rees, Minson & Smith, 1974; Ozanne *et al.,* 1976).

The estimation of the parameters needed for the development of models which can be used to synthesise optimal strategies may best commence with the construction of tentative models and the use of sensitivity tests to guide the allocation of research priorities. Undoubtedly, animals and estimates of animal production are likely to be necessary in the evaluation of responses, although long-term studies of animal production systems may neither be necessary nor feasible.

The usual type of stocking rate $\times$ fertilizer experiments may still play a key role. They can act as indicators of the changes which may take place in such pasture attributes as botanical composition, and as checks on predictions which arise from models. Extrapolations from such experiments should be made with caution, as the estimated response surfaces are valid only to the site and seasons that provided the results.

An optimal strategy might examine ways of using current information to predict animal responses. For example, in a mediterranean environment, if the autumn rains arrive early with sufficient following rains, there may be little point in applying fertilizer. In such circumstances pasture growth is likely to be sufficient so that additional growth would contribute only to surpluses. If rains arrived at about an average date the extra growth could be useful, and fertilizer application therefore worthwhile. But if the rains did not occur until mid-winter, fertilizer would be unlikely to alleviate significantly a feed shortage at least until early spring.

Additional information on pasture responses may be obtained from soil tests and from pasture response based on test strips or plots. Although probably not very accurate, these will have the advantage of being up to date and not subject to cumulative errors from previous decisions. The calibrations of pasture responses in terms of animal production, together with the economic interpretation of the inputs and immediate and long-term outputs, are the final steps in fertilizer evaluation, but physical estimation of responses will be feasible only on very few sites.

VIII. CONSERVATION AND SUPPLEMENTARY FEEDING

Hutchinson (1971) followed Macfadyen (1963) in regarding energy flow as a unifying concept for studying and comparing the productivity of systems. The argument is based on the fact that the systems which have the highest rates of production are, in a steady state, those which have the highest rates of respiration loss. This is clearly true since in a steady state, respiration equals photosynthesis. However, it must be qualified since the objective of agricultural systems is not to channel energy *through* the system but to channel energy (and protein, minerals, etc.) *into* harvestable and marketable material which should be appropriately weighted for quality, price and other measurements of value.

Supplementary feeding may affect grassland productivity if it permits higher stocking rates so that higher proportions of the total primary production are utilised. Supplementary feeding can be a short-term provision against regular feed shortages, which may be useful to bring animals to a suitable marketable condition, to enhance reproductive efficiency, and to avoid metabolic diseases such as pregnancy toxaemia, or it can be a long-term insurance against drought.

As Hutchinson (1971) and Bishop and Birrell (1975) emphasised, the need for supplementation must increase as stocking rates increase, but the opportunities for producing feed within the system will decrease at the same time. If one considers only the energy turnover, there will be some compromise stocking rate : conservation activity which will maximise the flow of energy through the domestic animals rather than through the ' decomposers ' of the system. If the system is wool production, or even meat production, the stocking rate which maximises the objective function probably will not maximise the energy flow through animals, but is likely to be substantially lower.

The use of fertilizers (especially nitrogen) may, by raising yields per ha, offer advantages through reducing the area to be conserved. Fodder crops may also provide useful feed at critical times and also be used for conservation of grain or of surpluses at other times. They can be grown, however, only at the expense of heavier stocking on the uncropped areas during those portions of the year in which the area reserved for the fodder crop is not available for grazing. The measurement of the effects of supplementary feeding and fodder conservation on grassland productivity must take into account management decisions such as where the feed comes from, when to feed, which animals to feed and how much to feed. These decisions can be critical and must either be justified by outside information, or be varied in investigations of supplementation and conservation systems.

Detailed examination of the many feasible variations one might introduce into the study of fodder-conservation systems would be inappropriate here. The review of Hutchinson (1971) is particularly valuable in describing the feedbacks in a closed system. The paper by Blaxter and Wilson (1963) indicates kinds of alternative strategies which may be visualised; the papers of Afzal, McCoy and Orazem (1965) and Morley and Graham (1971) provide examples of approaches which might be used in determining policies for conservation and provision of reserves against drought.

A system of fodder reserves and supplementary feeding may be based wholly or partly on fodder brought in from outside, or fodder may be produced and sold outside the system. Some systems may be completely closed, all reserves being produced within the system and used only within the system. The reserves may be destined to be held until needed during a drought, fire or flood, or other emergency, or they may be for supplementation during a regular seasonal need, such as during winter in cold climates. If they are produced within the system, production may or may not be from specified areas. These and other areas, may be part of the system of grassland production and should be clearly defined at the start of any investigation.

Livestock themselves can substitute the use of one resource for another. If given supplements they will modify their intake of pasture to an extent which depends on the type of animal, the supplement, and the pasture (Holder, 1962b; Leaver, Campling & Holmes, 1968; Umoh & Holmes, 1974; Newton & Young, 1974; Drew & Reid, 1975). Holder (1962) defined a coefficient of efficiency of supplementation which expressed the ratio of herbage consumed by supplemented animals to that consumed by non-supplemented animals. He found this ratio to be 0·57 for oat grain and Newton and Young (1974) obtained a ratio of 0·52 for barley and dried grass. Values such as these may be specific for the supplements and the pasture in question. Low quality herbage and more palatable supplements seem likely to decrease the efficiency of supplementation. The effects of supplementation on pasture availability at critical times is one possible consequence of supplementation for pasture productivity which has scarcely been studied.

The value of a supplement to pasture productivity may also depend on compensatory gains, reviewed by Wilson and Osbourn (1960) and Allden (1970). Studies of the energetics of restriction followed by plenty have shown that the overall use of energy is most efficient on a steady state (Keenan, McManus & Freer, 1969, 1970) and, although food intake was only slightly increased, if at all, the greater gains in weight were due largely to higher water contents in the gut, and in the regained tissues following restriction. In the field, studies, such as those of Meyer *et al.* (1965), Vivian (1970), Bennett *et al.* (1970), Scales and Lewis (1971) and others, demonstrated compensatory gains but there have been no clear indications that energetic efficiency was increased above that to be expected on a steady-state diet. The question of energy flow and conversion seems to remain open, and deserves further investigations such as those by Drew and Reid (1975 a, b) and Donnelly and Freer (1974).

Compensatory gains may affect grassland productivity by making fuller use of pasture when it is plentiful. Animal production in absolute terms, or per unit of intake, may not be increased if stock are marketed at the same time. If stock are marketed at equal weights, the restricted stock may have to be carried for a longer period. If this extra period is one of pasture surplus, extra productivity will probably have been achieved. If it is one of pasture deficit there may be no gains, but it may raise the question of whether to use the available supplementary feed in the middle of the fattening period or at the end. Carcase composition and grade, price differentials at different marketing times, the genotype of animals for sale, and the fodder and cash flows of the system may all intrude in the measurement of productivity and evaluation of supplements.

An investigation of conservation and supplementary feeding systems might compare the following variables :

(i) Type of material conserved—hay, silage, grain or straw from crops or pastures.
(ii) Level of conservation—the proportion of the area devoted to conservation or the number of food units per animal to be saved.
(iii) Rules for feeding—feeding may be practised only at critical times, or be based on allowing liveweight to fall below certain levels. In addition, the conserved material may be used either only to maintain animals, or to produce specified weight gains.
(iv) Use of aftermath—this must vary according to the system of production.
(v) Rules for evaluating, including decisions on marketing such as whether to market at fixed weights or at fixed times, and also the evaluation of surpluses or deficits of fodder.

Obviously, only a small fraction of the total number of possible systems can be evaluated. Therefore, the experimenter must set priorities as carefully as he is able in the light of existing information.

These complexities underline the necessity for clear definition of the production system in question, and careful appraisal of investigations on partial systems to ensure that feedback does not invalidate the conclusions.

The need for such studies is further emphasised by the results of Bailey and Bishop (1975) who found that periods of supplementary feeding markedly influenced the intake of cattle during subsequent grazing. The importance of changes in body composition in responses to changes in diet, as observed by Donnelly, Davidson and Freer (1974), also indicate difficulties in interpreting measurements such as liveweight change.

IX. SPECIAL PURPOSE PASTURES AND FODDER CROPS

Different grasslands or pasture plants may have different patterns of growth at different times of the year. For example, if moisture is available lucerne will grow more rapidly in summer than will ryegrass, but more slowly in the cooler months. An area of lucerne for grazing might, therefore, be included in a pasture system to meet special summer needs (e.g. green feed for weaners) and to make better use of summer rains in the belief that the advantages gained would outweigh the disadvantages of the lower production in the cool months.

Fodder cropping is frequently based on similar principles and expectations. The amount of pasture available during summer might be reduced in order to grow root crops for winter use. A cereal crop may be grown during summer for green feed in winter. Alternatively, it may be grown in winter on water stored by summer fallowing. It may then be fed to stock in winter and finally allowed to grow on for grain production.

The objectives underlying the use of special purpose pastures and fodder crops have been described by Wheeler (1968) as :

(i) To provide for special needs by techniques which are cheaper than supplementary feeding. For example, additional feed may be needed to meet the requirements for ewes at lambing, for cows at calving, or for weaners immediately after weaning.

(ii) To increase the use of basic resources by growing selected plants at times when plant production may be surplus to animal needs, and then harvesting them through grazing animals when special needs arise.

Myers (1967) discussed the theoretical allocation of areas and stocking rates at different times of the year for different pastures and proposed an experimental approach to the comparison of systems. The basic objective of his systems was to support a maximum number of animals on a fixed area throughout the grazing year. He emphasised that full experimental comparisons among all possible combinations even of as few as three pastures would scarcely be possible. His analysis demonstrated the importance of the carrying capacity of the special purpose pasture during the ' off peak ' periods as well as during the peak utilization period.

The most important unknown in setting up special-purpose systems seems likely to be the productivity (including carrying capacity) of each pasture type for each season. This might be estimated by means of tests with different stocking rates at different times of the year. Especially for extreme values, provision might have to be made to allow for carry-over effects in animals and to evaluate plant residues remaining from one period to another. It would indeed be sanguine to hope that optimal systems could be precisely defined. Fortunately, perhaps, optima are unlikely to be precisely definable since, as Myers has stated :

> " The penalties for not achieving this ideal may not be serious . . . The animal is not wholly dependent on current pasture growth. Excess pasture is carried over from one period to another . . . and animals can make compensatory gains in times of food plenty . . . As well, there are times when the animals can tolerate short-term shortages in diet without serious loss of production."

Unfortunately, there are also times when animals are unable to tolerate short-term shortages in their diet. These are predominantly around the time of parturition, mating, weaning (in weaners), and perhaps marketing and usually in dairy cattle; hence, it might be justified to incur substantial costs to meet such needs.

The cost of providing for special needs by means of special-purpose pastures or crops is difficult to define. A whole-system approach is usually necessary, but this may include situations where animals are brought into a system for only a brief period. For example, animals may be purchased in early winter, fattened on a crop and sold in early spring to meet a premium market.

It is necessary to take into account the total costs of growing and feeding the fodder crop to the whole system, as well as the benefits to animal production and the income from sale of the products of the crop. One of the most important costs, and one often overlooked when evaluating fodder cropping systems, is derived from the increases in stocking rates which are necessary on grazed pastures when the special-purpose pasture or crop land is not available for grazing. With fodder crops, if x percent of an area is used, the stocking rate on the remainder is increased by a factor of x/(100-x). The cost, although not obvious, might be substantial, being manifested in decreases in wool production, lower liveweights at joining, and lower weight gains by fattening stock; if, however, the additional stress is applied to cows during early pregnancy, it may cost relatively little. It may in some circumstances be even greater than the more obvious costs of land preparation, sowing and fertilizers.

The costs of the increased stocking rates may offset the advantages of resting special-purpose pastures. For example, in breeding females the extra nutritional stress would reduce ovulation rates and fertility at joining time and, therefore, reproductive performance also. The possible losses at parturition and during lactation could be avoided by using the special-purpose pasture or crop for its intended purpose. The costs of feeding the special-purpose pasture, although perhaps significant, seem unlikely to be critical. Reduced grain yields could be a major cost of grazing in some years, and in the years that grazing benefits grain yields, the use of the crop may not benefit the animals. This cost would depend on the price of grain, and would be substantial if grain prices were high and grain yields reduced (see Axelsen, Morley & Crouch, 1970).

An adequate experiment embracing stocking rates, fertilizer rates and other management variables on crops is unlikely to be feasible. Experiments might therefore be best oriented to estimating the most important parameters of appropriate models of systems. These seem likely to be :

- (i) the effect of increased stocking rate on uncropped areas on production during periods when crops are being grown or special-purpose pastures treated leniently;
- (ii) the amount of fodder available in the total system during the whole year;

(iii) the impact of levels of feed availability at various times on animal performance;

(iv) the effect of grazing, including perhaps different techniques of grazing, on subsequent crop yields;

(v) the crop yields obtained in a system aimed at crop production without the crop necessarily being grazed during the growing period and some evaluation of crop residues for animal production;

(vi) the response of fodder yields at different times of the year to variations in agronomic techniques.

Since all these parameters may vary greatly depending on the season, estimates derived from models based on meteorological data must be used for the synthesis of optimal systems. Such models should include variations in the proportion of farms which is devoted to the special-purpose crops or pastures.

The benefits obtained from the use of special-purpose pastures should, if feasible, be compared with those obtained from growing the important species together in one pasture. Sod-seeding of forage crops into pasture is an analogous activity. Stability of botanical composition might be aided by the use of planting, for example, alternate rows of lucerne and grass. The relative contributions from

PLATE 13. Cattle grazing under longleaf pine (*Pinus palustris*) in southwest Louisiana (Sternitzke & Pearson, 1975) (photo: US Forest Service)

different species could be regulated by variation in the numbers of rows or perhaps by row spacings, especially where perennials are sown into a base of annuals.

The use of crops for grazing is a special case of crop/animal integration. Nearly all agronomic studies concentrate on a single form of production and must therefore be of limited value to mixed enterprises. The study of wheat and sheep production by Cannon (1974) is a noteworthy advance, but much more information is needed on the value of stubbles and on the use of different kinds of crop for breeding and fattening livestock.

The integration of grazing with timber production is attracting increasing attention in North America (e.g. McLean, 1972; Pearson & Whitaker, 1974; Skovlin & Harris, 1974; Sternitzke & Pearson, 1975; see also Plate 13). Such systems have been in use for centuries, for example, in the *Quercus suber* and *Q. ilex* forests of Spain, where sheep, cattle, goats, pigs, cork bark, and revenues from hunting may all contribute to farm incomes, as well as to stability of enterprises.

X. SUBDIVISION OF GRASSLANDS

Subdivision is necessary for the grazing management of livestock, but whether subdivision of an area grazed by a given group is necessary, or even desirable, is by no means clear. Evidence based on animal production data is especially lacking, most experiments finding no advantage and some even disadvantages. The protagonists of subdivision can always fall back on the arguments that the grazing cycle was too long or too short, or that insufficient subdivisions were used. There is an inexhaustible supply of such rationalizations, some of which Heady (1961) has discussed.

The investigator must, therefore, design experiments which are, as far as possible, invulnerable to such arguments. It is not sufficient to compare, for example, rotational with continuous grazing. Rather, the main question studied should concern the optimum level of subdivision. Subsidiary questions might concern lengths of grazing periods, intervals between grazing, and interactions with other management practices. As more variables are considered, designs become unwieldy, if not impracticable.

The use of subdivision in grazing management may influence animal production in at least five ways :

(i) Pasture may be rationed so that more is available when it is most needed.
(ii) The growth of pasture may be increased so that more is available for animal consumption.
(iii) The botanical composition may be modified, so that the proportion of more nutritious species and parts of plants is increased or decreased.
(iv) Fouling by excreta may be reduced.
(v) The transmission of parasites and diseases may be reduced.

Unless these and other effects are clearly identified, any generalization must be restricted. Therefore investigations should be planned so that the causes of the differences in animal production can be identified, and this frequently poses difficult problems.

There are reasons for doubting whether pasture saving could ever be worthwhile. Firstly, the benefits obtained may not compensate for the effects of the restrictions which deferment imposes on the area available for grazing. Secondly, saved pasture is likely to produce less usable dry matter than an unsaved pasture, since leaf area indices might well be above optimum and rates of decay higher during much of the period of deferment (Hunt, 1965). Thirdly, and this may be most important, decisions to move livestock will usually be based on the quantity of feed available at any time. A farmer may cause decreases in animal production if he grazes subdivisions for too long, and hence too closely, in order to maximize 'utilization' of the pasture on one field while fodder goes to waste by decay on others. 'Percent utilization' of pastures has often been regarded as an important component of profit, but increased understanding of the concepts of optimum stocking rates has reduced, though not eliminated, emphasis on this as an objective. It may still, however, be responsible for decisions being based on dubious grounds.

Seasonal variability also creates uncertainty. If the season is favourable, feed is ample and deferment offers no real advantage. In bad seasons, the extra growth and feed available on deferred areas may be insufficient to confer much benefit when it is needed. Evaluation of strategies of rationing must therefore entail models which incorporate seasonal variations. That described by Freer *et al.* (1970) and extended by Christian *et al.* (1972, 1974) provides an example of preliminary developments. Experiments to provide information for, and to check, these models are also necessary. Such experiments will be needed over many years at a few sites or over a few years at many sites.

The production of dry matter by a pasture could be influenced by the use of subdivision. The mechanisms involved might act through variations in leaf area index, losses from decay, and the amount of chemical or morphological reserves available for regrowth. The growth rate may also be affected indirectly through changes in botanical composition, especially when these result in large changes in the ratio of summer to winter growing species.

It is sometimes argued that continuous grazing is similar to rotational grazing for the individual tiller. Hodgson (1966) and Greenwood and Arnold (1968) found that any given tiller was grazed at intervals which depend on grazing pressure, and also that sheep tended to eat the younger portions of leaves and to leave the older portions. The steady removal of younger leaves under continuous grazing could reduce the growth rate of the pasture more than would the almost complete removal of the herbage of all ages over the short periods of intensive grazing of rotationally grazed swards. If intake were limited by availability, set stocking could thus result in decreased animal production, but whether this ever actually happens because the dry matter production by the pasture is more limiting under one system or another is far from clear. The geometrical structure of the pasture (that is, leaf inclination and leaf area index) could be important, but again this is as yet obscure.

Cutting experiments such as that of Anslow (1967), or grazing in combination with cutting (Brougham, 1959), provide substantial evidence that yields of cut material are increased by longer intervals between cutting and grazing. Brougham (1959) also showed that ryegrass, white clover and red clover responded in different ways to harvesting intensity, the productivity of white clover being favoured by close grazing, that of red clover by lax grazing, and ryegrass interacting markedly with season. He also showed that grazing management affects the genotypic structure of a ryegrass sward (Brougham, Glenday & Fejer, 1960).

The effects of subdivision and grazing management on the botanical composition of pastures, evident in many studies, have been reviewed by Moore and Biddiscombe (1964). They range from modifications of the proportions of different species, especially clover and grass, through to the disappearance of species which were formerly major components of the sward. The mechanisms causing such changes probably vary greatly within and between species and may also depend on the properties of competing species. We are here concerned with the direct effects changes in botanical composition may have on animal production. These will depend on the properties and the proportions of the diet contributed by each species in the sward. Such contributions and their effects may vary with the time of year, especially if, for example, the pasture contains a mixture of warm-season and cool-season species.

Changes in botanical composition may affect animal production directly through changes in the quality and quantity of the diet and through the presence of toxins, e.g. alkaloids, isoflavones, of harmful plant structures, e.g. burrs, spines, awns, etc., and of substances which taint the meat or milk, and will depend on the type of livestock and the time of the year. For example, the loss of *Medicago sativa* or other legumes from a pasture may adversely affect young lambs or calves more than it affects adult animals which are better able to cope with poor quality feed. *Medicago hispida* may improve the plane of nutrition but its burrs may adversely affect the value of the fleece.

Wilson (1969) has indicated possible methods of investigating the grazing of browse plants, and the same principles apply to other forms of grassland. The presence of a particular component at high frequencies in the diet does not necessarily mean that changes in that component will change the nutritive value of the diet. Thus, Leigh, Wilson and Mulham (1968) found that although the shrub *Kochia aphylla* was a regular, and presumed important, component of the diet of sheep, its removal from the pasture had no discernible affect on animal production.

The losses which may be caused through fouling of pastures have been discussed by Marsh and Campling (1970), who emphasised the importance of grazing pressure in determining the proportion of fouled herbage to be rejected by the grazing animals. While increasing the stocking rate may increase the utilization of pasture, productivity may not be improved. Fouled pasture may become more acceptable after a few weeks but, according to Marten and Donker (1964, 1966), rejection was still quite strong even two months after the deposition of dung and therefore unlikely to be eliminated by subdivision and spelling pastures. According to Greenhalgh and Reid (1969) the consequences of these effects are not likely to be large with dairy cattle. Fluctuations in the nutrition of beef cattle and sheep may be less important and it thus seems doubtful if subdivision would contribute significantly to the reduction of losses associated with fouling of pastures (see also McKinney & Morley, 1975).

Since infective larvae and some other stages of the most important internal parasites develop slowly and survive for periods rather longer than pastures are usually spelled, the use of subdivision to reduce parasitic infection seems generally unlikely to be effective (e.g. Donald & Waller, 1973; Donald, 1974). There can be substantial reduction in internal parasitism if sheep and cattle can graze alternately on pastures for periods long enough to remove, or greatly reduce, each other's contamination. The advantages of this should be considered in relation to possible gains from mixed grazing, as shown by Hamilton and Bath (1970), and Bennett *et al.* (1970).

If they are to impart real understanding, experimental designs for the investigation of the effect of subdivision on grassland productivity must be much more complex than the ' rotational *v.* continuous grazing ' comparisons which have been usual. The results of Morley, Bennett and McKinney (1969) indicated strong non-linearity of results from 1, 3 and 9 subdivisions, with differences in botanical composition achieving critical importance. As Morley (1968b) suggested, there might not be much change in outcome of subdivision beyond 9 paddocks. The number of subdivisions investigated might conveniently be either 1, 2, 4 and 8, or 1, 3 and 9, or 1, 4 and 16, which would then cover the range from set stocking to an intensive system. If numbers such as these are chosen, stock movements can be synchronized across treatments, which may be of considerable benefit in operation and interpretation.

In addition to numbers of subdivisions, the length of the grazing cycle may be varied. If conservation levels or other variables are also imposed, an experiment can very rapidly become much too large and complex. Some investigators attempt to meet such difficulties by using incomplete factorials or single factorials without replications. The theoretical merits of these alternatives are generally well understood, and need not be elaborated here.

The principles underlying the study of the effects of subdivision can be clarified best by an example. The object of a study might be to estimate the effects of subdivision of a certain kind of pasture on animal production and on the pasture. First, one must decide whether the basis is to be on the number of paddocks per treatment, or on the lengths of grazing and spelling between grazings. We will assume that cycle lengths greater than 10 weeks are too long because too much pasture could decay, and that those shorter than 6 weeks are unlikely to be distinct from set stocking.

If the number of subdivisions is to be 3, 6 or 9 the following possibilities may be considered. The unit of time is taken to be one week.

No. Paddocks	*Grazing Period*	*Spelling Period*	*Cycle Length*	*Suitability*
3	1	2	3	?
3	2	4	6	+
3	3	6	9	+*
3	4	8	12	?
3	5	10	15	?
3	6	12	18	?
6	1	5	6	+
6	2	10	12	?
6	3	15	18	?
9	1	8	9	+*
9	2	16	18	?

Those combinations marked + under ' Suitability ' qualify as possible designs because of the length of the cycle. There are only four, which scarcely cover an adequate range of grazing times and spelling times. Therefore, the results obtained will not indicate whether differences obtained were due to differences in the length of the grazing period, the spelling period, or the whole cycle. A similar problem confronts the experimenter who wishes to use 1, 4, 6 and 8 paddock systems. Note that Morley, Bennett and McKinney (1969) used those treatments marked with an asterisk under ' Suitability ', and were unable to establish the cause(s) of the differences observed in terms of the lengths of the component periods of the system.

It may be more informative to study the effects of the lengths of the grazing period, the spelling period, and the grazing cycle. The following possibilities might be considered :

Grazing Period	*Spelling Period*	*Cycle Length*	*No. Paddocks*	*Suitability*
1	5	6	6	+
1	6	7	7	+
1	7	8	8	+
1	8	9	9	+
2	4	6	3	+
2	6	8	4	+
2	8	10	5	+

Grazing Period	*Spelling Period*	*Cycle Length*	*No. Paddocks*	*Suitability*
3	3	6	2	+
3	6	9	3	+
3	9	12	4	?
4	4	8	2	+
4	8	12	3	?

The ten treatments marked + under ' Suitability ', plus set stocking, would explore the region of interest fairly thoroughly. But the comparison of 11 production systems would present formidable problems. For example, there would be 77 paddocks in each replicate and this does not take into account the need to vary stocking rate.

It might be preferable to study the effects of spelling times and grazing times in small plots of different size. The stocking rate should be adjusted to be proportional to the ratio of the cycle length to the grazing period, and changes in swards could be studied in detail if there were no need to measure animal production in whole systems.

At the same time, production systems such as set stocking, one week grazing and 4 weeks spelling, 1 week grazing and 8 weeks spelling and 4 weeks grazing and 4 weeks spelling might be run to check and calibrate predictions from the small and intensive experiments and the models generated from them. The treatments chosen should provide real contrasts. It is unrealistically optimistic to hope that differences between 2-4, 2-6 or 3-6 treatments (grazing period/spelling period) are likely to be demonstrated in a finite experiment. The history of studies of the effects of subdivision, reviewed by Wheeler (1962), is replete with experiments from which unequivocal conclusions would be impossible because treatments are confounded, and replication often non-existent.

The experimenter should endeavour to understand the long-term consequences of grazing management on the pasture before he attempts detailed evaluation of production systems. Thus, it might be more useful, and much more feasible, to understand the consequences of grazing of various durations and intensities on botanical composition, soil structure, loss and runoff, than to obtain estimates of the production from some actual systems which are site- and year-specific and of little value for generalization.

Present information indicates that for most mesophytic pastures subdivision itself has little effect on pasture stability or animal production (Moore, Barrie & Kipps, 1946; Morley, Bennett & McKinney, 1969; Boswell *et al.,* 1974; Castle & Watson, 1973; Hood & Balsiie, 1973; Southcott *et al.,* 1972). The most important exception is lucerne, which does not persist under set-stocking except perhaps at very low stocking rates (Moore, Barrie & Kipps, 1946; McKinney, 1974; Fitzgerald, 1974; and others). This may be true of other temperate species such as *Trifolium pratense*, but evidence is lacking.

Most grazing experiments with tropical species have been set-stocked and this does not seem to have greatly affected animal production or the persistence of grasses and legumes ('t Mannetje, Jones & Stobbs, 1976). At least, with temperate or tropical species, failure of species to persist should be followed up by investigations of effects of grazing management.

In arid and semi-arid areas and in natural pastures generally, the importance of subdivision, pasture deferment and related techniques remains obscure. Undoubtedly, grazing management can markedly influence botanical composition, animal production, and pasture stability in regions such as Texas (Leithead, 1974) and in Southern Africa. Methods of management, variously termed 'Non-selective grazing', 'Short-duration grazing' and 'High intensity/low frequency grazing' have been used to describe systems claimed to be successful. These require further study, but at present it seems clear that results from highly improved pastures or from many annual pastures should not be applied uncritically to a wide range of pastures and environments.

XI. SYSTEMS, MODELS, GRASSLANDS AND ANIMAL PRODUCTION

Throughout this chapter it has become increasingly evident that although an impressive amount of research on animal production from grasslands has been completed, the use of this information in decision-making is beset with uncertainties of several kinds. Site, paddock, farm and regional variations inhibit extrapolation from research results to farmer decisions. Year-to-year variations in weather and in economic parameters introduce further difficulties. Moreover, the low accuracy of estimates of the effects of any treatment often allows real benefits to be overlooked and procedures of doubtful value to be accepted. This inaccuracy is the outcome not only of the inherent variability in the outputs from a production system, but also of the fact that only a small sample of possible treatment combinations can be studied.

Interactions between components in the production system add further complexities to decision-making. For example, subdivision or additional fertilizer may be beneficial at high stocking rates, but not at low. Supplementary feed may be worthwhile if a high proportion of ewes is bearing twins, and this proportion will be the outcome of pasture growth prior to and during joining. Interaction between components of systems is an inherent property of systems of animal production from grasslands.

This complexity has led many researchers in grassland systems to the techniques of systems analysis. A full discussion of systems analysis would be inappropriate here, but a general outline is attempted as an introduction to the concepts. A symposium (Australian Society of Animal Production, 1972), a book (Dent & Anderson, 1971) and a review (Anderson, 1974) are recommended as introductions to the literature. Models of production systems include those of Vickery and Hedges (1972), d'Aquino (1974), Arnold *et al.* (1974), Smith and Williams (1973), Bartlett, Evans and Bement (1974) and Edelsten and Newton (1975). Models by Noy-Meir (1975a, b) are of special interest.

The concepts of systems analysis are probably best indicated by an outline of procedures. These are usually as follows :

(i) Describe the system (e.g. beef production from natural pastures on the pampas of Argentina).
(ii) State the objective(s) (e.g., maximum profits with minimum soil losses).
(iii) Identify the problem (e.g., estimation of optimum stocking rate).
(iv) Identify the components (e.g., weather, pasture growth, animal intake, animal performance, sale value).
(v) Construct a mathematical model from available information and understanding, or relationships between these components (e.g., animal intake and pasture available).
(vi) Obtain predictions by running the model (simulation). If possible check predictions against available information (validation).
(vii) If predictions are unacceptable, try to identify the sources of inaccuracies, modify the model and proceed again.
(viii) Vary the control variables (e.g., stocking rates) of the model to answer the question in terms of the objective function. In this case the optimum stocking rate would be estimated by comparing profits and soil losses over a range of stocking rates, seasons and sites.

Most successful models have been developed for specific purposes. Reported failures often are due to ill-defined objectives and, equally, to the misapplication of an existing model. Biological systems are so complex that a full description is impossible. However, where one or two factors exert the major influence on system performance, very simple models can have powerful explanatory value (e.g. Reid & Thomas, 1973). The development of working models of agricultural production systems demands constant and close interaction between the conceptual and programming aspects of model development and purposeful field experimentation. Powerful and productive interactions between computer simulation and field and laboratory experimentation can and should be expected in the systems mode of research.

The consequences for agricultural research of such interactions include the following :

* Precise definition of objectives.
* Assembly of available information into a logical framework.
* Identification of important gaps in understanding.
* Indication of priorities for research.
* Use of information to aid decision making at all levels.

* Greater relevance of research to present or potential agricultural production systems.
* Indication of potential opportunities for increased efficiency prior to physical investigation. Physical investigations of such opportunities might then proceed purposefully in the light of estimates of benefits which may be achieved.

Some risks which may be associated with extensive involvement in model development should be recognised and guarded against. These include :

* Preoccupation with modelling and with computer technology as ends in themselves rather than as aids to decision making and extrapolation;
* Emphasis on those parts of systems that are most amenable to mathematical information, but which may not be major determinants of the efficiency of agricultural systems;
* Distraction of attention from field experimentation and the acquisition of solidly based information upon which the validity of models depends;
* Neglect of research which seeks to understand processes, develop new chemical compounds, improve varieties, or develop radically different forms of production. That is, there should be a balance between research on systems and research on processes.

Mathematical modelling and field experimentation may both be valid forms of systems research when pursued independently. The greatest gains in agricultural research will arise from the combination of the two, as this should lead to the maximum of generalization from the data from field experiments. Field experiments give results which are largely specific to one site, one period of time, and a limited set of treatments. Mathematical models seek to generalize to many sites and years, and to a wider range of treatments. Obviously, the efficiency of this generalization must depend to a large extent on the design of the experiments and on the data which are collected.

The semi-diagrammatic table on p. 145 is an attempt to summarize some of the more important relationships between components of animal production systems.

These have been selected from reports in the literature, especially those of Wright (1970), Vickery and Hedges (1972) and Australian Society of Animal Production (1972). The shapes of response curves have certainly not been established for the great majority of relationships, and in most cases experimental investigation is unlikely to do so in the near future. The curves are no more than approximate descriptions which seem likely to apply over a considerable range of associations.

These relationships may be considered as primary functions from which other relationships can be derived. The relationship between stocking rate and reproductive performance, for example, may be derived from

SUMMARY OF THE MORE IMPORTANT RELATIONSHIPS BETWEEN COMPONENTS OF ANIMAL PRODUCTION SYSTEMS

DEPENDENT VARIABLES

INDEPENDENT VARIABLES AND PROBABLE RESPONSES

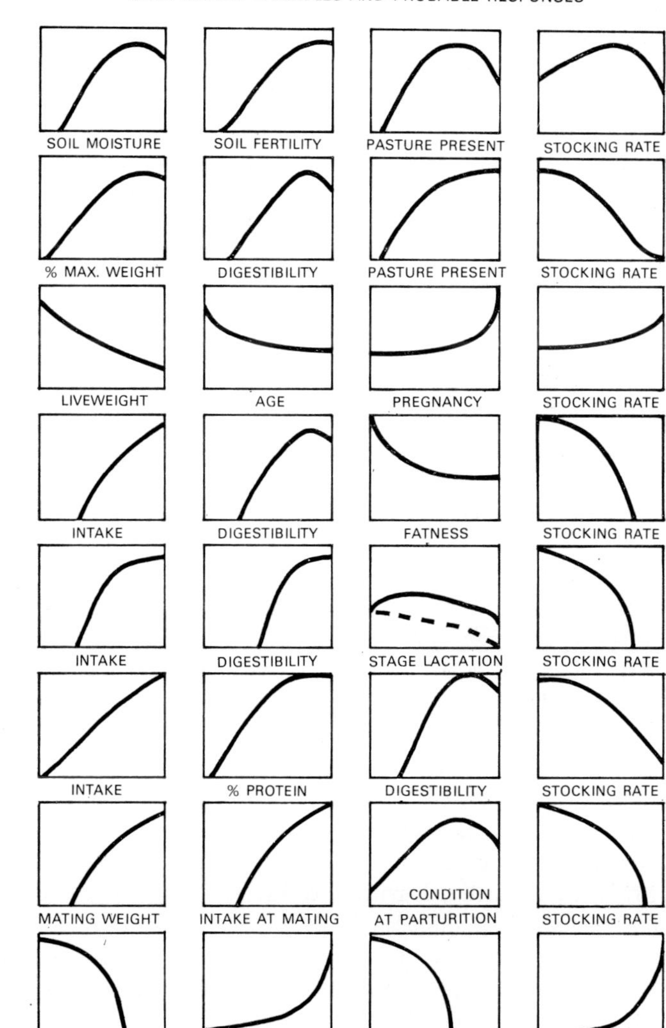

the amount of pasture present, the growth of pasture, the intake of animals, the metabolic needs of animals, the liveweight gains, the amount of supplementary feeding, the weight at mating, the intake at mating, and the condition at parturition.

The curves shown indicate that pasture growth approaches a maximum by a sigmoid pathway for soil moisture and temperature. The response to amount of pasture present may be of similar shape, or it may be a ' diminishing returns ' response such as that to soil fertility and perhaps light energy. High temperatures, waterlogging, high concentration of salts in the soil and large quantities of pasture all reduce pasture growth. The maximum growth rate, and the level of temperature, moisture, etc. at which growth rate is maximum, will vary among plant species and with the structure of the pasture.

The pasture intake of animals responds rapidly to the amount of pasture available, up to levels at which the animal's appetite is quickly satisfied and other limitations such as digestibility come into force. The slope of the curve is likely to differ appreciably among different pasture species. Below certain levels of digestibility the intake is likely to approach a maximum for the pasture in question. Intake may even decrease in animals which have become obese.

The metabolic requirement of animals is a function of metabolic size, usually estimated as $W^{0.75}$. The appropriate exponent to use for different animal species seems to remain open to further examination (Graham, 1972). Metabolic needs decrease per unit weight as animals mature, especially during the first year of life of both sheep and cattle.

Needs will increase in pregnant animals as parturition approaches. In early pregnancy fairly severe nutritional stress may be sustained without important penalties, although some embryo loss may be induced shortly after fertilization (Cumming, 1972a, b) and lamb growth may be adversely affected (Curll, Davidson & Freer, 1975). A high level of fertility demands high inputs of energy prior to and during mating. There are suggestions that low levels of pasture availability may increase the metabolic cost of grazing, but this cost is by no means precisely defined.

Liveweight gains vary greatly in composition and energy content, depending on whether the animal is, and has been, gaining or losing weight (Drew & Reid, 1975a, b; Donnelly, Davidson & Freer, 1974). The composition of gains is fairly standard for animals of a given size and maturity (Donnelly & Freer, 1974) so that the energy cost of gains is a simple function of the energy costs of storage of the components, plus the maintenance requirements of the animal. This is the usual basis for predicting gains in grazing models. The intake of energy is a function of the amount of pasture present and its digestibility, the size and age of the animal, and certain little-understood attributes of pastures (e.g. palatability, legumes or grasses).

Responses of milk production may be somewhat dampened by the extent to which energy is used either for milk or for liveweight gains. These will vary with the genotype of the animal, the stage of lactation and perhaps with the quality of the ration.

Wool production may depend substantially on the amount and quality of protein ingested. Therefore, a response curve is not suggested here. For a ration of a given quality, the production of wool is probably linear with digestible organic matter intake (Ferguson, 1972).

The proportion of females which become pregnant will vary with the nutritional status prior to and during mating. This is obviously of critical importance to animal production. The performance during lactation will also be related to the nutritional status both before and after parturition. Difficulties at parturition and excess lactation may accompany high planes of nutrition prior to parturition, but quantitative relationships have not been clearly established.

The relationship between mortality at any time and body condition may be critically important in animal production and necessary for the construction of decision functions for supplementary feeding. Liveweight could be deceptive, because variations in the water content in tissues can mask variations in energy reserves. Changes in mortality associated with age may also be important in studies of different age structure of flocks and herds.

Investigations of these relationships in the context of production systems are certain to be relevant to the estimation of animal production from grasslands. The relative importance of any one of them must be evaluated by sensitivity tests before research priorities can be allocated, keeping in mind also the relative feasibility and costs of research and the probability of success. The inaccuracies and ignorance of any of the relationships used to synthesise or evaluate systems must be recognised. For example, it would be unrealistic to expect that one curve could describe the relationship between intake and pasture availability for any class of animal over all types of grassland. But at present the hard facts necessary to construct better curves are not plentiful.

Relationships must often be used which are the product of informed guesswork, intuition and reasoning from basic principles, but if they incorporate the best available information, they should provide the best predictions. The alternative, to let formulation of decision functions await the provision of precisely documented data, is scarcely acceptable.

Variables that have a binomial distribution, such as fertility and mortality, pose special problems. The estimation of the effects of management on fertility in a field experiment with livestock would involve very large numbers of animals if it is to be so accurate that an economically important difference is statistically evident. For example, the standard error of an estimate of fertility of 0·6 from 100 individuals would be approximately 0·049. The 90 percent confidence interval would thus lie between 0·5 and 0·7, values which have substantially different implications for an enterprise.

It is therefore debatable whether attempts to obtain precise estimates of treatment effects on binomially distributed variables can be justified. The effects of treatments may well be estimated from concomitant variables with sufficient precision to be useful. For example, the effects of a treatment on the fertility of cows or ewes may perhaps be better estimated from knowledge of the relationship between fertility and liveweight at mating, rather than from the results of an experiment with 50 females in each treatment. The use of only sufficient animals to measure continuously distributed variables, such as liveweight, might release resources to allow primary relationships to be studied in greater depth so that they may then contribute information of value over a wide range of enterprises.

The interpretation of relationships such as that between fertility and liveweight can be misleading. For example, a simple regression of fertility and liveweight (e.g. Coop, 1962), especially in genetically heterogeneous flocks, could be the sum of genetic and environmental influences. The regression coefficient would not, therefore, indicate the response which might be obtained from supplementation. An appropriate form of analysis of this relationship in cattle is presented by Morley, Axelsen and Cunningham (1976).

XII. MEASUREMENT OF HERBAGE IN LARGE GRAZING EXPERIMENTS

Although measurement of herbage in general has been covered in Chapter 4, large grazing experiments pose particular problems which will be outlined here.

The interpretation of animal production data from grazing experiments depends on estimates of the amount of components of the pasture eaten, the amounts present, and the nutritional value of the components. From this information an experimenter can construct a picture of the flow of food and energy in the production system. This may indicate where limitations to production may be present, or seem likely to be present, and such information may in turn lead to suggestions for overcoming limitations.

The inclusion of frequent measurements of pastures in grazing experiments seems only seldom to have importantly aided interpretation of results. Indeed, few experimenters seem to consider that such results merit publication, except in rough outline. This appears to be justified where the observations are made at times when pasture quantity or quality quite clearly do, or do not, limit animal performance. If however animal performance is below (or above) expectation, one might ask a specific question such as " is performance limited by pasture availability, or digestibility or some other attribute ?" One might then attempt to answer this question by estimations on the pasture, and perhaps on the animals. The resources used in this attempt then stand a good chance of being spent usefully.

If pasture measurements are made at specified intervals without particular regard to animal needs or performance, much of the effort may well represent inefficient use of resources. Therefore, a close watch on animal performance which permits unexpected results to be translated almost instantly into specific questions, the questions into estimates, and the estimates into understanding of phenomena which may still be present, could add greatly to the interest of grazing experiments and to the information obtained from them.

The measurement of herbage in field experiments presents three types of problems :

1. Sampling variation due to :
   (a) heterogeneity of site and differential grazing within plots (camping areas, nearness to water, etc.);
   (b) variations over short distances due to differential grazing, botanical composition, and localised variations in fertility, water relations, etc.
2. Errors of measurement due to :
   (a) observers or equipment (bias);
   (b) precision of observers or equipment (consistency).
3. Logistic problems arising from relationships between the area to be covered and the equipment, observers and time available.

   In general the number of observers is limited, and the observations, whether mechanical, visual or electronic, must be compressed into a relatively short interval. A short interval is necessary to avoid drifts in time and to accommodate measurements within the total programme.

1. Sampling variations

Macro-variation between large portions of the site may best be overcome by systematic stratification of the plots so that if n measurements are made, each 1/nth of each plot receives one measurement. In practice an approximation may be adopted by roughly dividing a plot into some number, say m, areas ($m < n$) and taking n/m measurements in each area. There are several ways of doing this, of which the use of markers such as fence posts and the following of routes marked on a map of the plot are examples.

Random sampling of a plot can never achieve the same precision per measurement as stratified sampling, but it will approach the same precision as the value of n becomes large. Moreover, the achievement of truly random sampling in the field presents substantial problems. The locations could be chosen by some suitable random sampling technique (e.g. co-ordinates on a map) but this would present problems of orientation in the field. But if such problems can be overcome the opportunity exists to make stratification more precise.

Micro-variation over distances of a few centimetres or a metre or two can seldom be systematically sampled except in very small plots. There are many local variations of techniques of achieving randomness, and probably most are open to abuse by observers who may consciously or unconsciously select some preferred type of herbage.

2. Errors of measurement and logistics

Bias due to observers or equipment can, and must, be corrected by suitable calibration. The techniques for calibration of equipment such as electronic probes are well understood. The calibration of observers is similar in principle, but presents additional problems. Fatigue and the immediate past experience (e.g. the nature of the previous plot observed) of the observer should be taken into account, if possible.

Training of observers could offer important benefits if visual or other subjective assessments are to be used. In practice when using observers with some acquaintance with pastures, the fact that they have a lot of previous experience has not resulted in obvious benefits to precision. The attitude of observers to the job may be much more important. Observers should be physically fit so that they are not inhibited by the not insubstantial physical efforts needed. They should understand the importance of their observations so that they are motivated to maintain mental acuity despite physical fatigue.

The precision of individual measurements may be quite low but this can be compensated for by increasing the number of measurements and by the use of such techniques as ranking. It may sometimes be preferable to increase the number of measurements rather than to use techniques such as ranking. This will depend on the additional precision achieved by ranking relative to the cost in time, fatigue or other limitations, and the additional stratification which may accompany increased numbers of estimates. Ranking may achieve most in relatively small plots of heterogeneous herbage, but least in large plots with much macro-variation. These arguments suggest the following important requirements in planning large-scale experiments :

(i) Careful selection and briefing of observers, especially if subjective estimates are to be made. In practice such selection is often not possible.

(ii) Planning so that observers can complete their measurements without undue exertion and within the time available.

(iii) Calibration extending throughout the period of measurement to take account of changes in time, in instruments, the pasture, or the observers.

XIII. RELEVANCE OF RESEARCH RESULTS TO ANIMAL PRODUCTION SYSTEMS

As discussed by Morley and Spedding (1968), research results may be of limited relevance to real production systems unless the experiment represented a complete production system or a reasonably self-contained portion of such a system. This is not to say that all measurements of animal production from grasslands should have some pragmatic objective or be oriented to some production system; it is simply a question of the relevance of the results. Perhaps many investigators consider that the main objective of biologists is to obtain biological information, and that virtually all such information is bound to be useful. Unfortunately, this objective cannot be considered adequate; certainly it is often doubtful whether the most important questions have been asked and asked in the right way.

Because grazing experiments are expensive in time and money, the relevance and importance of the kinds of results expected should be assured before details of planning are commenced. For example, Morley and Spedding (1968) paid particular attention to ' put and take ' methods of experimentation and questioned the appropriateness of the results of such experiments to grassland evaluation. The arguments have not been accepted by some writers. For example, Matches (1970) has defended the ' put and take ' method, although he admits that " farmers and ranchers cannot apply the practice of varying animal numbers ". He contends that alternate procedures are applicable and cites conservation of excess forage during flush periods, subdivision and deferment, and supplementation in times of scarcity. No doubt, such practices can be incorporated into many production systems but unless they can be justified, the interpretation of results is jeopardized. Arguments such as : " . . . the most powerful aspect of the ' put and take ' method is that when grazing pressure is maintained at definable levels throughout the season, the quality and quantity time trends of the sward can be identified . . . . " cannot be accepted. The quality and quantity time trends identified might be quite irrelevant to the production system(s) in question unless measurements were made in the context of such a system. Matches goes on to suggest that " . . . as researchers we should be less concerned as to the merits of the ' put and take ' system and more concerned with defining plant/animal interrelationships ". This viewpoint seems to neglect the fact than an infinity of relationships might be defined and studied but their relevance will depend on the occurrence of such relationships in actual production systems.

There are at least two production systems in which ' put and take ' techniques may be perfectly valid. This may be so in a system where the majority of animals are in a feed-lot and pasture is a relatively minor component of the particular production system, although perhaps of prime importance to the whole industry. Variation of the stocking rate in such a way as to obtain maximum benefit from the pasture may well be a normal part of the production system and, therefore, appropriate in experiments concerned with grain producing farms. This may apply to areas similar to the Corn Belt in the USA. Secondly, on farms where most pastures are relatively poor but are being improved, the production system on the farm consists of the animals, and the areas of improved and unimproved pastures. The former area will usually be superior in respect to both quality and quantity. It would then be perfectly valid to vary the stocking rate on the improved area in order to maximize production from the total system. The work of Cotsell (1956) is an outstanding example of this approach. He studied systems in which the improved pastures on the whole area were rationed to different classes of stock by permitting access for varying periods. For example, he studied the effects of 3·5 or 7 days per week on the improved pastures for breeding ewes, zero and 3·5 days per week for young sheep, and zero, 2 or 5 days per week for dry sheep.

The development of optimal strategies for the use of limited areas of improved or special-purpose pastures by such means remains a high priority where pasture improvement is still a major activity. ' Put and take ' techniques could well be the main method of evaluation, rather than the techniques, used by most investigators, that are appropriate to wholly improved areas. But it is most important that the ' put and take ' on some pastures, and ' take and put ' on others, were evaluated as total systems. Apart from Cotsell, few, if any, workers seem to have asked such important questions in the developmental stages of pasture improvement. The fact that his research had largely to be self-supporting is probably relevant since it led directly to a definition of problems of pasture use and evaluations.

If a production system is being studied, the yield of a particular variant of the system is best and most simply expressed in terms of the objective of the system. As a corollary, the productivity of a system can validly be expressed only in relation to the system used. Thus Burton (1970) stated that annual liveweight gains of 390, 470 and 500 kg/ha were obtained when pastures fertilized with 100 kg/ha of N were stocked at 3·3, 4·4 or 5·5 steers per ha. He also stated that when fertilized with 225 kg N/ha, liveweight gains of 765 kg/ha were obtained during the grazing season. His facts are not questioned, but when interpreting these facts it is important to keep in mind that the latter yield was obtained by the use of a ' put and take ' system, carrying up to 11 steers per ha. The first of his statements might well be a useful index of the productivity of the grassland in question in a total system. The second statement cannot be taken as an index of productivity because the liveweight gains achieved by some of the animals used in the tests were obtained only by the use of resources outside the pasture considered. The first statement may be a useful guide to the value of the species and management used; the second could be grossly misleading except in relation to a particular system which should be precisely defined so that its relevance to a real farm situation can be judged.

The use of ' put and take ' systems invokes problems in computing yields of animal product because of the difficulties of evaluating the performances of different kinds of testers. The most sophisticated estimates involve the use of feeding standards in reverse in order to calculate the number of ' effective feed units ' (EFU) harvested (Petersen & Lucas, 1968). It can be argued that many, if not all, of these difficulties disappear if one is considering the productivity of the total grassland system, and if the objective of the system has been precisely defined.

Wheeler *et al.* (1973) have discussed the choice of fixed or variable stocking rates at some length. Their arguments seem to be met by the principles enunciated above, and it seems unfortunate that so many of the experiments reported recently have used designs which make the results of dubious value for on-farm decision-making.

The development and improvement of production systems depends on the plant-animal relationships which are pertinent to that system, and the pertinence of a relationship may best be established by studying physical or conceptual models of actual or possible systems. One might obtain a profound understanding of relationships of little or no relevance to animal production. That may be a very worthwhile activity for some purposes, but it should not claim to be concerned with improvement of agricultural productivity.

XIV. CONCLUSIONS

This chapter has considered the measurement of grassland productivity through animal production largely from the viewpoint of decision-making on the farm, rather than from that of the research biologist. Economic criteria must be introduced into the processes of evaluation. A case has been made for studying grassland productivity in the context of production systems. Measurements on animal production are always necessary at some stages, but not at all. Insistence of the use of animal production throughout in procedures of grassland evaluation could be both wasteful and ineffective, although evaluation without fairly rigorous tests by animal production at any stage could be misleading. The need for efficient experimental designs is emphasised by the costs of research in this field. These costs also underline the need to be quite clear as to what kind of information is to be obtained, and how this information is likely to be used.

The testing of management practices should be designed within the framework of the decisions that have to be taken and the value of the information gained in decision-making. Since decision-making must be a dynamic procedure which takes account of new information as it becomes available with the progress of time and the accumulation of experience, the evaluation and development of management practices should be based on increasingly precise definition of criteria for decisions, and better estimates of those criteria. This could involve the development of dynamic decision functions and also quite radical departures from conventional field experimentation.

This study has emphasised the need for a systems approach in grassland evaluation, and some aspects of model development are therefore discussed. It is obvious that few relationships between relevant variables are known with a precision sufficient to justify confidence in extrapolating from research results to field practice. Since such precision is not likely to be obtained soon, the need for models becomes increasingly apparent. Even if imperfect, they can use all the information which is available. The evaluation of grassland productivity is quite clearly an exercise in decision-making under conditions of uncertainty. It is likely to remain so for an indefinite period, but relevant research should reduce the uncertainties as well as open the doors to new opportunities.

XV. REFERENCES

ACOCKS, J. P. H. (1966). Non-selective grazing as a means of veld reclamation. *Proc. Grassl. Soc. S. Afr.* **1,** 33-9.

AFZAL, M., MCCOY, J. H. & ORAZEM, F. (1965). Development of inventory models to determine feed reserves for beef-cattle production under unstable climatic conditions. *J. Farm. Econ.* **47,** 948-62.

ALLDEN, W. G. (1970). The effects of nutritional deprivation on the subsequent productivity of sheep and cattle. *Nutr. Abstr. Rev.* **40,** 1167-84.

ANDERSON, J. R. (1974). Simulation: Methodology and application in agricultural economics. *Rev. Market. Agric. Econ.* **42,** 3-55.

ANSLOW, R. C. (1967). Frequency of cutting and sward production. *J. Agric. Sci.* **68,** 377-84.

D'AQUINO, S. A. (1974). A case study for optimal allocation of range resources. *J. Range Mgmt.* **27,** 228-33.

ARNOLD, G. W., BALL, J., MCMANUS, W. R. & BUSH, I. G. (1966). Studies on the diet of the grazing animal. I. Seasonal changes in the diet of sheep grazing on pastures of different availability and composition. *Aust. J. Agric. Res.* **17,** 543-56.

ARNOLD, G. W., CARBON, B. A., GALBRAITH, K. A. & BIDDISCOMBE, E. F. (1974). Use of a simulation model to assess the effects of a grazing management on pasture and animal production. *Proc. XIIth Int. Grassl. Congr. Sect. Pap. 'Grassland Utilization'*, pp. 47-52.

AUSTRALIAN SOCIETY OF ANIMAL PRODUCTION (1972). Systems Analysis Symposium. *Proc. Aust. Soc. Anim. Prod.* **9,** 1-138.

AXELSEN, A. & MORLEY, F. (1968). Evaluation of eight pastures by animal production. *Proc. Aust. Soc. Anim. Prod.* **7,** 92-8.

AXELSEN, A., MORLEY, F. H. W. & CROUCH, M. (1970). Comparisons of sheep and grain production from systems of pasture and oats. *Proc. Aust. Soc. Anim. Prod.* **8,** 422-7.

BAILEY, P. J. & BISHOP, A. H. (1975). Liveweight change, grazing time of steers and effect of pasture height on liveweight change following periods of hand feeding. *Aust. J. Exp. Agric. Anim. Husb.* **15,** 440-5.

BARROW, N. J. (1967). Some aspects of the effects of grazing on the nutrition of pastures *J. Aust. Inst. Agric. Sci.* **33,** 254-62.

BARTLETT, E. T., EVANS, G. R. & BEMENT, R. E. (1974). A serial optimization model for ranch management. *J. Range Mgmt.* **27,** 233-9.

BENNETT, D., MORLEY, F. H. W., CLARK, K. W. & DUDZINSKI, M. L. (1970). The effect of grazing cattle and sheep together. *Aust. J. Exp. Agric. Anim. Husb.* **10,** 694-709.

BISHOP, A. H. & BIRDELL, H. A. (1975). Efficiency of grazing/fodder-conservation systems. *Proc. IIIrd World Conf. Anim. Prod.*, pp. 217-25.

BLACKBURN, A. G., FREW, M. V. & MULLANEY, P. D. (1973). Estimating optimum economic stocking rates for wethers. *J. Aust. Inst. Agric. Sci.* **39,** 18-23.

BLAXTER, K. L. & WILSON, R. S. (1963). The assessment of a crop husbandry technique in terms of animal production. *Anim. Prod.* **5,** 27-42.

BOSWELL, C. C., MONTEATH, M. A., ROUND-TURNER, N. L., LEWIS, K. H. C. & CULLEN, N. A. (1974). Intensive lamb production under continuous and rotational grazing systems. *N.Z. J. Exp. Agric.* **2,** 403-8.

BROCKMAN, J. S., SHAW, P. G. & WOLTON, K. M. (1970). The effect of phosphate and potash fertilizers on cut and grazed grassland. *J. Agric. Sci.* **74,** 397-407.

BROUGHAM, R. W. (1959). The effects of frequency and intensity of grazing on the productivity of a pasture of short-rotation ryegrass and red and white clover. *N.Z. J. Agric. Res.* **2,** 1232-48.

BROUGHAM, R. W., GLENDAY, A. C. & FEJER, S. O. (1960). The effects of frequency and intensity of grazing on the genotypic structure of a ryegrass population. *N.Z. J. Agric. Res.* **3.** 442-53.

BROWNE, D. (1966). Nitrogen use on grassland. 1. Effect of applied nitrogen on animal production from a ley. *Ir. J. Agric. Res.* **5,** 89-102.

BROWNE, D. (1967). Nitrogen use on grassland. 2. Effects of applied nitrogen on animal production from an old permanent pasture. *Ir. J. Agric. Res.* **6,** 73-81.

BROWNE, D. & WALSHE, M.J. (1968). Nitrogen use on grassland. 3. Effects of nitrogen and stocking rate on production per animal and per acre. *Ir. J. Agric. Res.* **7,** 121-8.

BRYAN, W. W. & EVANS, T. R. (1973). Effects of soils, fertilizers and stocking rates on pastures and beef production on the Wallum of south-eastern Queensland. I. Botanical composition and chemical effects on plants and soils. *Aust. J. Exp. Agric. Anim. Husb.* **13,** 516-29.

BURTON, G. W. (1970). Symposium on pasture methods for maximum production in beef cattle: Breeding and managing new grasses to maximise beef cattle production in the south. *J. Anim. Sci.* **30,** 143-7.

BUTLER, G. W. (1975). Effect of grazing systems on health and productivity. *Proc. IIIrd World Conf. Anim. Prod.,* pp. 174-9.

BYRNE, P. F. (1968). An evaluation of different stocking rates of merino sheep. *Bur. Agric. Econ. (Aust.) Res. Rep. No.* 13, (Mimeo., 80 pp.)

CAMPBELL, A. G. (1966a). Effects of treading by dairy cows on pasture production and botanical structure, on a Te Kowhai soil. *N.Z. J. Agric. Res.* **9,** 1009-24.

CAMPBELL, A. G. (1966b). Grazed pasture parameters. I. Pasture dry-matter production and availability in a stocking rate and grazing management experiment with dairy cows. *J. Agric. Sci.* **67,** 199-210.

CAMPBELL, A. G. (1966c). Grazed pasture parameters. II. Pasture dry-matter use in a stocking rate and grazing management experiment with dairy cows. *J. Agric. Sci.,* **67,** 211-6.

CAMPBELL, A. G. (1966d). Grazed pasture parameters. III. Relationships of pasture and animal parameters in, and general discussion of a stocking rate and grazing management experiment with dairy cows. *J. Agric. Sci.,* **67,** 217-21.

CANNON, D. J. (1972). The influence of rate of stocking and application of superphosphate on the production and quality of wool, and on the gross margin from Merino wethers. *Aust. J. Exp. Agric. Anim. Husb.* **12,** 348-54.

CANNON, D. J. (1974). Integrating sheep and wheat production: effect of substituting wheat production for pasture on the wool production of wethers. *Aust. J. Exp. Agric. Anim. Husb.* **14,** 454-60.

CARTER, E. D. & DAY, H. R. (1970). Interrelationships of stocking rate and superphosphate rate on pasture as determinants of animal production. I. Continuously grazed old pasture land. *Aust. J. Agric. Res.* **21,** 473-91.

CASTLE, M. E. & WATSON, J. N. (1973). A comparison between a paddock system and a " Wye College " system of grazing for milk production. *J. Br. Grassl. Soc.* **28,** 7-11.

CHISHOLM, A. H. (1965). Towards the determination of optimum stocking rates in the high rainfall zone. *Rev. Marketing Agric. Econ.* **33,** 5-31.

CHRISTIAN, K. R., ARMSTRONG, J. S., DAVIDSON, J. L., DONNELLY, J. R. & FREER, M. (1974). A model for decision-making in grazing management. *Proc. XIIth Int. Grassl. Congr. Sect. Pap. ' Grassland utilization '*, pp. 126-31.

CHRISTIAN, K. R., ARMSTRONG, J. S., DONNELLY, J. R., DAVIDSON, J. L. & FREER, M. (1972). Optimization of a grazing management system. *Proc. Aust. Soc. Anim. Prod.* **9,** 124-9.

CONNIFFE, D., BROWNE, D. & WALSHE, M. J. (1970). Experimental design for grazing trials. *J. Agric. Sci.* **74,** 339-42.

CONNIFFE, D., BROWNE, D. & WALSHE, M. J. (1972). An example of a method of statistical analysis of a grazing experiment. *J. Agric. Sci.* **79,** 165-7.

COOP, I. E. (1962). Liveweight-productivity relationships in sheep. I. Liveweight and reproduction. *N.Z. J. Agric. Res.* **5,** 249-64.

COTSELL, J. C. (1956). Sheep investigations at Shannon Vale nutrition station with special reference to strategic stocking. *Proc. Aust. Soc. Anim. Prod.* **1,** 24-32.

CUMMING, I. A. (1972a). The effects of increasing and decreasing liveweight on ovulation and embryonic survival in the Border Leicester × Merino ewe. *Proc. Aust. Soc. Anim. Prod.* **9,** 192-8.

CUMMING, I. A. (1972b). The effect of nutritional restriction on embryonic survival during the first three weeks of pregnancy in the Perendale ewe. *Proc. Aust. Soc. Anim. Prod.* **9,** 199-203.

CURLL, M. L., DAVIDSON, J. L. & FREER, M. (1975). Efficiency of lamb production in relation to the weight of the ewe at mating and during pregnancy. *Aust. J. Agric. Res.* **26,** 553-65.

DAVIES, H. L. & GREENWOOD, E. A. N. (1972). Liveweight and wool growth of sheep in response to the quantity and botanical composition of annual pasture induced by nitrogen fertilizer and stocking treatments. *Aust. J. Agric. Res.* **23,** 1101-11

DAVIES, H. L., GREENWOOD, E. A. N. & WATSON, E. R. (1966). The effect of nitrogenous fertilizer on wool production and liveweight of merino wether sheep in south western Australia. *Proc. Aust. Soc. Anim. Prod.* **6,** 222-8.

DAVIS, A. W. & HALL, W. B. (1969). Cyclic change-over designs. *Biometrika* **56,** 283-93.

DENT, J. B. & ANDERSON, J. R. (Eds.) (1971). Systems analysis in agricultural management. Sydney: John Wiley & Son.

DONALD, A. D. (1974). Some recent advances in the epidemiology and control of helminth infection in sheep. *Proc. Aust. Soc. Anim. Prod.* **10,** 148-55.

DONALD, A. D. & WALLER, P. F. (1973). Gastro-intestinal nematode parasite populations in ewes and lambs and the origin and time course of infective larval availability in pastures. *Int. J. Parasitol.* **3,** 219-33.

DONNELLY, J. R., AXELSEN, A. & MORLEY, F. H.W. (1970). Effect of flock size and grazing management on sheep production. *Aust. J. Exp. Agric. Anim. Husb.* **10,** 271-8.

DONNELLY, J. R., DAVIDSON, J. L. & FREER, M. (1974). Effect of body condition on the intake of food by mature sheep. *Aust. J. Agric. Res.* **25,** 813-23.

DONNELLY, J. R. & FREER, M. (1974). Prediction of body composition in live sheep. *Aust. J. Agric. Res.* **25,** 825-34.

DREW, K. R. & REID, J. T. (1975a). Compensatory growth in immature sheep. I. The effects of weight loss and realimentation on the whole body composition. *J. Agric. Sci.* **85,** 193-204.

DREW, K. R. & REID, J. T. (1975b). Compensatory growth in immature sheep. II. Some changes in the physical and chemical composition of sheep half-carcass following feed restriction and realimentation. *J. Agric. Sci.* **85,** 205-20.

EBERSOHN, J. P. (1966). Effects of stocking rate, grazing method and ratio of cattle to sheep on animal liveweight gains in a semi-arid environment. *Proc. Xth Int. Grassl. Cong.* pp. 495-9.

EDELSTEN, P. R. & NEWTON, J. R. (1975). A simulation model of intensive lamb production from grass. *Grassl. Res. Inst., (Hurley). Tech. Rep.*

EDMOND, D. B. (1966). The influence of animal treading on pasture growth. *Proc. Xth Int. Grassland Cong.* pp. 453-8.

EDMOND, D. B. (1970). Effects of treading on pastures, using different animals and soils. *Proc. XIth Int. Grassl. Congr.* pp. 604-8.

FERGUSON, K. A. (1972). The nutritional value of diets for wool growth. *Proc. Aust. Soc. Anim. Prod.* **9,** 314-20.

FIRTH, J. A., EVANS, T. R. & BRYAN, W. W. (1975). Effects of soils, fertilizers and stocking rates on pastures and beef production on the Wallum of south-east Queensland. 4. Budgetary appraisals of fertilizer and stocking rates. *Aust. J. Exp. Agric. Anim. Husb.* **15,** 531-40.

FITZGERALD, R. D. (1974). The effect of intensity of rotational grazing on lucerne density and ewe performance at Wagga Wagga, Australia. *Proc. XIIth Int. Grassl. Congr. Sect. Pap. ' Grassland utilization '*, pp. 180-5.

FREER, M., DAVIDSON, J. L., ARMSTRONG, J. S. & DONNELLY, J. R. (1970). Simulation of summer grazing. *Proc. XIth Int. Grassl. Cong.* pp. 913-7.

GRAHAM, N. MCC. (1972). Units of metabolic body size for comparisons amongst adult sheep and cattle. *Proc. Aust. Soc. Anim. Prod.* **9,** 352-5.

GREENHALGH, J. F. D. & REID, G. W. (1969). The effects of grazing intensity on herbage consumption and animal production. III. Dairy cows grazed at two intensities on clean or contaminated pasture. *J. Agric. Sci.* **72,** 223-8.

GREENWOOD, E. A. N. & ARNOLD, G. W. (1968). The quantity and frequency of removal of herbage from an emerging annual grass sward by sheep in a set-stocked system of grazing. *J. Br. Grassl. Soc.* **23,** 144-8.

GREENWOOD, E. A. N., DAVIES, H. L. & WATSON, E. R. (1967). Growth of an annual pasture on virgin land in south-western Australia including effects of stocking rate and nitrogen fertilizer. *Aust. J. Agric. Res.* **18,** 447-59.

HAMILTON, D. & BATH, J. G. (1970). Performance of sheep and cattle grazed separately and together. *Aust J. Exp. Agric. Anim. Husb.* **10,** 19-26.

HART, R. H. (1972). Forage yield, stocking rate, and beef gains on pasture. *Herb. Abstr.* **42,** 345-53.

HEADY, E. O. & DILLON, J. L. (1961). Agricultural production functions. Iowa: State Univ. Press.

HEADY, H. F. (1961). Continuous vs. specialized grazing systems : A review and application to the California annual type. *J. Range Mgmt.* **14,** 182-93.

HEADY, H. F. (1970). Grazing systems : Terms and definitions. *J. Range Mgmt.* **23,** 59-61.

HENZELL, E. F. & STIRK, G. B. (1963a). Effects of nitrogen deficiency and soil moisture stress on growth of pasture grasses at Samford, south-east Queensland. 1. Results of field experiments. *Aust. J. Exp. Agric. Anim. Husb.* **3,** 300-6.

HENZELL, E. F. & STIRK, G. B. (1963b). Effects of nitrogen deficiency and soil moisture stress on growth of pasture grasses at Samford, south-east Queensland. 2. Calculation of the expected frequency of dry periods by a water-budget analysis. *Aust. J. Exp. Agric. Anim. Husb.*, **3,** 307-13.

HILDER E. J. (1964). The distribution of plant nutrients by sheep at pasture. *Proc. Aust. Soc. Anim. Prod.* **5,** 241-8.

HODGSON, J. (1966). The frequency of defoliation of individual tillers in a set-stocked sward. *J. Br. Grassl. Soc.* **21,** 258-63.
HOLDER, J. M. (1962). Supplementary feeding of grazing sheep—its effect on pasture intake. *Proc. Aust. Soc. Anim. Prod.* **4,** 154-9.
HOLMES, W. (1968). The use of nitrogen in the management of pasture for cattle. *Herb. Abstr.* **38,** 265-77.
HOOD, A. E. M. & BAILIE, J. H. (1973). A new grazing system for beef cattle—the two field system. *J. Br. Grassl. Soc.* **28,** 101-8.
HOOPER, J. F. & HEADY, H. F. (1970). An economic analysis of optimum rates of grazing in the California annual-type grassland. *J. Range Mgmt.* **23,** 307-11.
HUMPHREYS, L. R. (1966). Pasture defoliation practice : a review. *J. Aust. Inst. Agric. Sci.* **32,** 93-105.
HUNT, L. A. (1965). Some implications of death and decay in pasture production. *J. Br. Grassl. Soc.* **20,** 27-31.
HUTCHINSON, K. J. (1971). Productivity and energy flow in grazing/fodder conservation systems. *Herb. Abstr.* **41,** 1-10.
JAGTENBERG, W. D. (1970). Predicting the best time to apply nitrogen to grassland in spring. *J. Br. Grassl. Soc.* **25,** 266-71.
JARDINE, R. (1975). Two cheers for optimality! *J. Aust. Inst. Agric. Sci.* **41,** 30-4.
JARDINE, R., O'BRIEN, S. & FREW, M. V. (1975). Relationship of wool production to stocking rate in Victoria. *Aust. J. Exp. Agric. Anim. Husb.* **15,** 357-62.
JOBLIN, A. D. H., BLACKMORE, L. W., BIRCHAM, J. S., COSSENS, G. G., CUMBERLAND, G. L. B., O'CONNOR, M. B., THOMSON, N. A., SMITH, R. G. & WRIGHT, D. F. (1972). Review of field research section fertilizer × stocking rate grazing trials in New Zealand. *Proc. N.Z. Soc. Anim. Prod.* **32,** 64-76.
JONES, R. J. (1974). The relation of animal and pasture production to stocking rate on legume based and nitrogen fertilized subtropical pastures. *Proc. Aust. Soc. Anim. Prod.* **10,** 340-3.
JONES, R. J. & SANDLAND, R. L. (1974). The relation between animal gain and stocking rate. Derivation of the relation from the results of grazing trials. *J. Agric. Sci.* **83,** 335-42.
KEENAN, D. M., MCMANUS, W. R. & FREER, M. (1969). Changes in the body composition and efficiency of mature sheep during loss and regain of live weight. *J. Agric. Sci.,* **72,** 139-47.
KEENAN, D. M., MCMANUS, W. R. & FREER, M. (1970). Voluntary intake of food by mature sheep following restricted feeding. *J. Agric. Sci.,* **74,** 477-85.
LEAVER, J. D., CAMPLING, R. C. & HOLMES, W. (1968). Use of supplementary feeds for grazing dairy cows. *Dairy Sci. Abstr.* **30,** 355-61.
LEIGH, J. H., WILSON, A. D. & MULHAM, W. E. (1968). A study of merino sheep grazing a cotton-bush (*Kochia aphylla*)-grassland (*Stipa variabilis-Danthonia caespitosa*) community on the Riverine plain. *Aust. J. Agric. Res.* **19,** 947-61.
LEITHEAD, H. L. (1974). High intensity-low frequency grazing system increases livestock production. *Proc. XIIth Int. Grassl. Congr. Sect. Pap. 'Grassland utilization'* pp. 355-61
LESPERANCE, A. L., JENSEN, E. H., BOHMAN, V. R. & MADSEN, R. A. (1960). Measuring selective grazing with fistulated steers. *J. Dairy Sci.* **43,** 1615-22.
LUCAS, H. L. (1960). Critical features of good dairy feeding experiments. *J. Dairy Sci.* **43,** 193-212.
MCARTHUR, I. D. (1970). Optimum sheep stocking rate. *J. Aust. Inst. Agric. Sci.* **36,** 9-14.
MCARTHUR, I. D. & DILLON, J. L. (1971). Risk, utility and stocking rate. *Aust. J. Agric. Econ.* **15,** 20-35.

MACFADYEN, M. A. (1963). Animal ecology: Aims and methods. London: Pitman & Sons, 2nd Edn.

MCKINNEY, G. T. (1974). Management of lucerne for sheep grazing on the Southern Tablelands of New South Wales. *Aust. J. Exp. Agric. Anim. Husb.* **14,** 726-34.

MCKINNEY, G. T. & MORLEY, F. H. W. (1975). The agronomic role of introduced dung beetles in grazing systems. *J. Appl. Ecol.* **12.,** 831-7.

MCLEAN, A. (1972). Beef production on Lodgepole Pine—Pinegrass range in the Cariboo region of British Columbia. *J. Range Mgmt.* **25,** 10-1.

MCMEEKAN, C. P. (1960). Grazing management. *Proc. VIIIth Int. Grassl. Congr.* pp. 21-6.

MANNETJE, L. 't, JONES, R. J. & STOBBS, T. H. (1976). Pasture evaluation by grazing experiments. *In* Tropical pasture research. Principles and methods. Eds. N. H. Shaw & W. W. Bryan. *Commonw. Bur. Pastures Field Crops, Hurley. Bull.* 51, pp. 194-234.

MARSH, R. & CAMPLING, R. C. (1970). Fouling of pastures by dung. *Herb. Abstr.* **40,** 122-30.

MARSH, R. & MURDOCH, J. C. (1974). Effect of high fertilizer nitrogen and stocking rates on liveweight gain per animal and per hectare. *J. Br. Grassl. Soc.* **29,** 305-11.

MARTEN, G. C. & DONKER, J. D. (1964). Selective grazing induced by animal excreta. II. Investigation of a causal theory. *J. Dairy Sci.* **47,** 871-4.

MARTEN, G. C. & DONKER, J. D. (1966). Animal excrement as a factor influencing acceptability of grazed forage. *Proc. Xth Int. Grassl. Congr.* pp. 359-63.

MATCHES, A. G. (1970). Pasture research methods. *Proc. Nat. Conf. Forage Qual. Eval. Util., Univ. Nebr.* Sect. 1, pp. 1-32.

MEYER, J. H., HULL, J. L., WEITKAMP, W. H. & BONILLA, S. (1965). Compensatory growth responses of fattening steers following various low energy intake regimes on hay or irrigated pasture. *J. Anim. Sci.* **24,** 29-37.

MOORE, R. M. (1970). Australian grasslands. *In* Australian grasslands. Ed. R. M. Moore, Canberra: Australian National University Press, pp. 87-100.

MOORE, R. M., BARRIE, N. & KIPPS, E. H. (1946). Continuous and rotational grazing by merino sheep. 1. A study of the production of a sown pasture in the Australian Capital Territory under three systems of grazing management. *Aust. CSIR. Bull.* 201, pp. 7-69.

MOORE, R. M. & BIDDISCOMBE, E. F. (1964). The effects of grazing on grasslands. *In* Grasses and grasslands. Ed. C. Barnard. Melbourne: Macmillan & Co. Ltd., pp. 221-35.

MORLEY, F. H. W. (1962). Pasture plant breeding and animal production. *J. Aust. Inst. Agric. Sci.* **28,** 3-7.

MORLEY, F. H. W. (1968a). Economics of pasture improvement. *In* Pasture improvement in Australia. Ed. B. Wilson, Sydney: K. G. Murray Publishing Co., pp. 105-17.

MORLEY, F. H. W. (1968b). Pasture growth curves and grazing management. *Aust. J. Exp. Agric. Anim. Husb.* **8,** 39-45.

MORLEY, F. H. W. (1972). A systems approach to animal production. What is it about ? *Proc. Aust. Soc. Anim. Prod.* **9,** 1-9.

MORLEY, F. H. W. (1974). Evaluation by animal production of increases in pasture growth, using computer simulation. *Proc. XIIth Int. Grassl. Congr. Sect. Pap. ' Grassland utilization '*, pp. 416-9.

MORLEY, F. H. W., AXELSEN, A. & CUNNINGHAM, R. B. (1976). Liveweight at joining and fertility in beef cattle. *Proc. Aust. Soc. Anim. Prod.* **11,** 201–4.

MORLEY, F. H. W., BENNETT, D. & MCKINNEY, G. T. (1969). The effect of intensity of rotational grazing with breeding ewes on phalaris-subterranean clover pastures. *Aust. J. Exp. Agric. Anim. Husb.* **9,** 74-84.

MORLEY, F. H. W. & GRAHAM, G. Y. (1971). Fodder conservation for drought. *In* Systems analysis in agricultural management. Eds. J. B. Dent & J. R. Anderson. Sydney: John

Wiley & Sons, pp. 212-36.

MORLEY, F. H. W. & SPEDDING, C. R. W. (1968). Agricultural systems and grazing experiments. *Herb. Abstr.* **38,** 279-87.

MORRIS, R. M. (1969). The pattern of grazing in ' continuously ' grazed swards. *J. Br. Grassl. Soc.* **24,** 65-70.

MOTT, G. O. (1960). Grazing pressure and the measurement of pasture production. *Proc. VIIIth Int. Grassl. Congr.*, pp. 606-11.

MYERS, L. F. (1967). Assessment and integration of special purpose pastures. I. Theoretical. *Aust. J. Agric. Res.* **18,** 235-44.

MYERS, L. F. (1972). Effects of grazing and grazing systems. *In* Plants for sheep in Australia. Eds. J. H. Leigh & J. C. Noble. Sydney: Angus and Robertson, pp. 183-92.

NEWTON, J. E. & YOUNG, N. E. (1974). The performance and intake of weaned lambs grazing S24 perennial ryegrass with and without supplementation. *Anim. Prod.* **18,** 191-9.

NOY-MEIR, I. (1975a). Stability of grazing systems: an application of predator-prey graphs. *J. Ecol.* **63,** 459-81.

NOY-MEIR, I. (1975b). Rotational grazing in a continuously growing pasture: a simple model. *J. Agric. Syst.* **1,** 87–112.

OWEN, J. B. & RIDGMAN, W. J. (1968). The design and interpretation of experiments to study animal production from grazed pasture. *J. Agric. Sci.* **71,** 327-35.

OZANNE, P. G. & HOWES, K. M. W. (1971a). The effects of grazing on the phosphorus requirement of an annual pasture. *Aust. J. Agric. Res.* **22,** 81-92.

OZANNE, P. G. & HOWES, K. M. W. (1971b). Preference of grazing sheep for pasture of high phosphate content. *Aust. J. Agric. Res.* **22,** 941-50.

OZANNE, P. G., PURSER, D. B., HOWES, K. M. W. & SOUTHEY, I. (1976). Influence of phosphorus content on feed intake and weight gain in sheep. *Aust. J. Exp. Agric. Anim. Husb.* **16,** 353–60.

PATTERSON, H. D. & LUCAS, H. L. (1962). Change-over designs. *Agric. Exp. Stn. Tech. Bull. No.* 147.

PEARSON, H. A. & WHITAKER, L. B. (1974). Forage and cattle responses to different grazing intensities on Southern pine ridge. *J. Range Mgmt.* **27,** 444-6.

PETERSEN, R. G. & LUCAS, H. L. (1968). Computing methods for the evaluation of pastures by means of animal response. *Agron. J.* **60,** 682-7.

PETERSEN, R. G., LUCAS, H. L. & WOODHOUSE, W. W. (1956). The distribution of excreta by freely grazing cattle and its effect on pasture fertility. *Agron. J.* **48,** 440-9.

REED, K. F. M. (1972). The performance of sheep grazing different pasture types. *In* Plants for sheep in Australia. Eds. J. H. Leigh & J. C. Noble. Sydney: Angus and Robertson, pp. 192-204.

REES, M. C., MINSON, D. J. & SMITH, F. W. (1974). The effect of supplementary and fertilizer sulphur on voluntary intake, digestibility, retention time in the rumen, and site of digestion of pangola grass in sheep. *J. Agric. Sci.*, **82,** 419-22.

REID, G. K. R. & THOMAS, D. A. (1973). Pastoral production, stocking rate and seasonal conditions. *Q. Rev. Agric. Econ.* **26,** 217-27.

RIEWE, M. E. (1961). Use of the relationship of stocking rate to gain of cattle in an experimental design for grazing trials. *Agron. J.* **53,** 309-13.

ROBERTS, B. R. (1967). Non-selective grazing—an evaluation. *Farmers' Weekly* (*S. Afr.*) **113,** 37.

SANDLAND, R. L. & JONES, R. J. (1975). The relation between animal gain and stocking rate in grazing trials; an examination of published theoretical models. *J. Agric. Sci.* **85,** 123-8.

SCALES, G. H. & LEWIS, K. H. C. (1971). Compensatory growth in yearling beef cattle. *Proc. N.Z. Soc. Anim. Prod.* **31,** 51-61.

SCOTT, R. S. (1968). Animal and pasture production as indices of fertilizer maintenance requirements. *Proc. N.Z. Soc. Anim. Prod.* **28,** 53-64.

SHAW, N. H. (1970). The choice of stocking rate treatments as influenced by the expression of stocking rate. *Proc. XIth Int. Grassl. Congr.* pp. 909-13.

SKOVLIN, J. M. & HARRIS, R. W. (1974). Grassland and animal response to cattle grazing methods in the ponderosa pine zone. *Proc. XIIth Int. Grassl. Congr. Sect. Pap. ' Grassland utilization '*, pp. 639-45.

SMITH, R. C. G. & WILLIAMS, W. A. (1973). Model development for a deferred-grazing system. *J. Range Mgmt.* **26,** 454-60.

SMOLIAK, S. (1974). Range vegetation and sheep production at three stocking rates on *Stipa-Bouteloua* prairie. *J. Range Mgmt.* **27,** 23-6.

SOUTHCOTT, W. H., ROE, R. & NEWTON TURNER, H. (1962). Grazing management of native pastures in the New England region of New South Wales. II. The effect of size of flock on pasture and sheep production with special reference to internal parasites and grazing behaviour. *Aust. J. Agric. Res.* **13,** 880-93.

SOUTHCOTT, W. H., WHEELER, J. L., HILL, M. K. & HEDGES, D. A. (1972). Effect of subdivision, stocking rate, anthelmintic and selenium on the productivity of Hereford heifers. *Proc. Aust. Soc. Anim. Prod.* **9,** 408-11.

STERNITZKE, H. S. & PEARSON, H. A. (1975). Forest-range resources of southwest Louisiana *J. Range Mgmt.* **28,** 264-6.

STOBBS, T. H. & SANDLAND, R. L. (1972). The use of a latin square change-over design with dairy cows to detect differences in the quality of tropical pastures. *Aust. J. Exp. Agric. Anim. Husb.* **12,** 463-9.

TADMOR, N. H., EYAL, E. & BENJAMIN, R. W. (1974). Plant and sheep production on semiarid annual grassland in Israel. *J. Range Mgmt.* **27,** 427-32.

TILL, A. R. & MAY, P. F. (1973). Nutrient cycling in grazed pastures. *J. Anim. Sci.* **37,** 813-20.

UMOH, J. E. & HOLMES, W. (1974). The influence of type and level of supplementary feed on intake and performance of beef cattle on pasture. *J. Br. Grassl. Soc.* **29,** 301-4.

VICKERY, P. J. (1972). Grazing and net primary production of a temperate grassland. *J. Appl. Ecol.* **9,** 307-14.

VICKERY, P. J. & HEDGES, D. A. (1972). Mathematical relationships and computer routines for a productivity model of improved pasture grazed by Merino sheep. *Aust. CSIRO Anim. Res. Lab. Tech. Pap. No.* 4.

VIVIAN, G. F. (1970). The effects of two management practices on the growth of steers between eight and twenty months and on the composition of the grazed pasture. *Aust. J. Exp. Agric. Anim. Husb.* **10,** 137-44.

WALSHE, M. J. (1975). Grazing management and the productivity of grazing systems. *Proc. IIIrd World Conf. Anim. Prod.,* pp. 165-73.

WEIR, W. C. & TORELL, D. T. (1959). Selective grazing by sheep as shown by a comparison of the chemical composition of range and pasture forage obtained by hand clipping and that collected by esophageal-fistulated sheep. *J. Anim. Sci.* **18,** 641-9.

WHEELER, J. L. (1962). Experimentation in grazing management. *Herb. Abstr.* **32,** 1-7.

WHEELER, J. L. (1968). Major problems in winter grazing. *Herb. Abstr.* **38,** 11-18.

WHEELER, J. L., BURNS, J. C. MOCHRIE, R. D. & GROSS, H. D. (1973). The choice of fixed or variable stocking rates in grazing experiments. *Exp. Agric.* **9,** 289-302.

WILLOUGHBY, W. M. (1975). The evaluation of grazing systems. *Proc. IIIrd World Con. Anim. Prod.* Ed. R. L. Reid, Sydney: University Press, pp. 158-64.

WILSON, A. D. (1969). A review of browse in the nutrition of grazing animals. *J. Range Mgmt.* **22,** 23-8.

WILSON, P. N. & OSBOURN, D. F. (1960). Compensatory growth after undernutrition in mammals and birds. *Biol. Rev.* **35,** 324-63.

WOLFE, E. C. & LAZENBY, A. (1973). Grass-white clover relationships during pasture development. *Aust. J. Exp. Agric. Anim. Husb.* **13,** 567-80.

WRIGHT, A. (1970). Systems research and grazing systems management-oriented simulation. *Univ. N. Engl. Farm Mgmt. Bull. No.* 4.

XVI. GENERAL READING LIST

ALEXANDER, G. I. (Ed.) (1973). Manual of techniques for field investigations with beef cattle. Canberra, Australia: CSIRO.

BARNES, R. F., CLANTON, D. C., GORDON, C. H., KLOPFENSTEIN, T. J. & WALDS, D. R. (Eds.) (1970). *Proc. Nat. Conf. Forage Qual. Eval. Util., Univ. Nebraska.*

CAMPBELL, J. B. (1969). Experimental methods for evaluating herbage. *Can. Dept. Agric. Publ.* 1315.

DILLON, J. L. (1968). The analysis of response in crop and livestock production. London: Pergamon Press.

GOLLEY, F. B. & BUECHNER, H. K. (Eds.) (1968). A practical guide to the study of the productivity of large herbivores. Int. Biol. Program. Handbook 7. Oxford: Blackwell Scientific Publications.

GRASSLAND RESEARCH INSTITUTE, HURLEY (1961). Research techniques in use at the Grassland Research Institute, Hurley. *Commonw. Bur. Pastures Field Crops, Hurley, Berkshire, Bull.* 45.

HARLAN, J. R. (1956). Theory and dynamics of grassland agriculture. Princeton N.J.: Van Nostrand.

HEADY, H. F. (1967). Practices in range forage production. Brisbane: University of Queensland Press.

HENDERSON, C. R. (1959). Design and analysis of animal husbandry experiments. *In* Techniques and procedures in animal production research. Maryland: American Society of Animal Production, pp. 1-55.

MCCLUSKEY, D. S. (1959). Studies of herbage production and consumption and the production of dairy cows in various grazing conditions. *In* The measurement of grassland productivity, Ed. J. D. Ivins. London: Butterworths.

MOORE, R. M. (Ed.) (1970). Australian grasslands. Canberra: Australian National University Press.

MOULE, G. R. (Ed.) (1965). Field investigations with sheep: a manual of techniques. Melbourne: CSIRO.

SHAW, N. H. & BRYAN, W. W. (Eds.) (1976). Tropical pasture research. Principles and methods. *Commonw. Bur. Pastures Field Crops, Hurley, Berkshire, Bull.* 51.

TROPICAL GRASSLAND SOCIETY OF AUSTRALIA (1971). Limitations to dairy production in the tropics. Conference held at Wollongbar, N.S.W. *Trop. Grassl.* **5,** 145-302.

TROPICAL GRASSLAND SOLIETY OF AUSTRALIA (1973). The Mulga lands of Australia. Symposium held at Charleville, Queensland. *Trop. Grassl.* **7,** 1-170.

Chapter 7

MEASURING ANIMAL PERFORMANCE

J. L. Corbett

I. INTRODUCTION

All components of a grazing system interact and all must be considered if a real understanding of the system is to be achieved. Plant production is a function of climate, soil, the populations of microorganisms and invertebrates above and below the soil surface, and the activities of the larger animals whether feral or domesticated. Production from animals depends on the amount, quality and seasonal pattern of plant growth, on their efficiency in harvesting the herbage and in utilizing what they consume. In this chapter are discussed measurements that may be made on grazing animals, with particular reference to circumstances in which particular techniques can be applied and problems in their use.

Several types of measurement can be made fairly readily and, for example, liveweight gains or milk production as determined over a range of stocking densities (Chapter 6) will give information on the productivity over a period of time of the particular area of vegetation studied. More detailed studies are necessary to understand why the observed level of animal production is obtained and why it may vary within and between years, and the relevance of the information to other areas and situations. In some instances quite simple procedures may give insight into the processes involved; thus the determination of faecal output alone may sometimes provide an adequate index of herbage intake. Particular problems may require the use of detailed and complex procedures, though there must be reasonable assurance that the often large expenditure of resources is justified. Whatever their degree of complexity the procedures must be used in their correct context and with an appreciation of the errors that might be involved. For example, an equation of Watson and Horton (1936) for predicting " with a high error of estimation " the energy value (starch equivalent) of " grassland herbage with a fair proportion of clover " has been widely misused. Other examples include the uncritical adoption of equations for predicting the digestibility and intake of pasture from faecal composition.

II. PASTURES AND ANIMAL NUTRITION

The biology of grazing animals is similar to that of comparable animals hand-fed (McDonald, 1968), but there are characteristic differences between the respective diets. That from pasture or browse is highly variable in composition; at one extreme the feed available may have the energy and protein values of a concentrate feed and at the other extreme it may approach those of sawdust. The material eaten is chiefly leaves and stems, usually with relatively small amounts of seeds so that, unlike many concentrates, starch is generally only a minor component. Water content is highly variable, ranging from about 10 percent in dry standing herbage to more than 90 percent in growing herbage wet with dew or rain. Moreover, although grazing intake is theoretically *ad libitum* and is not amenable to fine control by man, the amounts actually eaten may be severely restricted if the quantity of feed available per unit area is small. In addition, within the limits of availability, the animal has free choice among the plants comprising the grazed community; these can include grasses, legumes and many other dicotyledonous species (forbs) and shrubs and trees (browse), and the animal may select particular parts of a plant species. These activities have large interactive effects on the grazed community. There will be direct and consequential effects of defoliation on the morphology and botanical composition of the plant cover and its micro-environment; changes in soil conditions, including temperature, moisture and distribution of plant nutrients owing to transfer in excreta; changes in soil bulk density, structure, water penetration and aeration owing to treading; and finally changes in the large and varied populations of micro-organisms and invertebrate animals below and above the soil surface.

Despite these differences between grazing and hand-feeding, the nutrients required by animals in the two situations are the same, but the quantities needed may differ. For example, the grazing animal is likely to require more energy for maintenance because generally it has to work harder to obtain feed and is more subject to inclement weather; however, the increases are less than have been indicated by some reports in which the requirement, estimated as grazing intake for constant liveweight, appeared to be several times greater than for comparable housed animals. Subsequent work showed that intakes were overestimated owing to bias from causes not then realized, and that there is no warrant for suggesting that the nutrition of grazing animals is in any way peculiar or inexplicable in terms of existing principles. This sometimes appears to be implied in references to ' luxury consumption ', i.e. herbage intake in excess of immediate requirements, even raising suspicions that in such circumstances the first law of thermodynamics might not hold!

From a nutritional standpoint the productivity of grazed vegetation is determined by its capacity to sustain the supply of energy and nutrients needed by the animal for maintenance, and production of the desired type. The production responses by the animal may be regarded as depending on its rate of feed intake(I) feed digestibility (D), and the efficiency of use of the metabolites from digestion (E) This factorization provides a convenient concept of nutritional value ($NV = I \times D \times E$), and of the productivity of an area of vegetation when grazed ($\Sigma_{i=1}^{n}[NV]$, where n is the number of animals). Studies can be made of each term separately and of relationships with types and conditions of pastures and animals, though the interactions of each with the others must always be borne in mind (Raymond, 1969; Corbett, 1969).

Productivity of ecosystems is frequently described in terms of energy flow. Similarly, the nutrition and productivity of grazing animals is frequently examined in terms of dietary energy intake and its utilization. " The basic need of animals fed normal rations is for energy and this demand is the basis for most, and perhaps all of the other nutrient requirements " (Crampton, 1956); and again " The energy need of the animal [is] a primary one to which requirements of other nutrients are linked " (Blaxter, 1967). The level of animal production sustained by pastures is most often determined by the energy content of the feed. Such problems as inadequacies or excesses of mineral nutrients are relatively local or temporary.* Increases in the amounts of amino acids available for absorption from the small intestine achieved by protecting proteins from digestion in the rumen may increase production at pasture, as in stall-feeding; these particular circumstances excepted, the nitrogen (N) content of the diet of grazing animals in temperate regions is in general commensurate with the energy content. In subtropical and tropical regions, inadequacies in N content occur more frequently, as with mature herbage during the dry season when the N content of the dry matter often falls below 1 percent; microbial activity

---

* Specific discussion of mineral nutrition, identification and amelioration of disorders, and of nutritional disorders generally (e.g. bloat, phytotoxicity) is outside the scope of the chapter. These topics and many others are discussed by Butler and Bailey (1973), Underwood (1966) and McDonald (1968).

in the rumen and the protein status of the animal (Egan, 1965) may then become impaired, causing a decline in feed intake. Provision of a supplement such as urea may sometimes promote greater feed intake and so improve animal performance, but many of the experiments on this topic illustrate, often by default, an important principle, namely that short-term studies of some isolated aspect of pastoral production may give results that are incomplete or misleading (see Chapter 6). If, for example, provision of urea for a restricted period does improve animal performance, the longer term benefits are doubtful because compensatory growth in unsupplemented animals during a subsequent period of good feed supplies will often quite overshadow the effects of the treatment (Loosli and McDonald, 1968).

III. SOME DEFINITIONS AND NUTRITIONAL INDICES

The discussion in this section is mainly on nutritional value per unit weight of feed, but probably the most important factor determining the level of animal production is the amount eaten. There is, in general, a positive relationship between the voluntary intake of feed by ruminant animals and its digestibility, though the relationship is neither constant nor exact. It is of less importance when digestibility exceeds about 65 percent than at lower values, and for grazing rather than hand-fed animals because the former tend to select the more digestible components of the pasture herbage (see 'Digestibility of Grazed Herbage' on p. 191).

1. *Digestibility*

The apparent digestibility of a feed, so called because matter of endogenous (body) origin in faeces is not distinguished from feed residues proper, is usually referred to simply as the digestibility (D) and as a percentage is :

$$D = \left\{ \frac{I - F}{I} \right\} \times 100 \qquad (1)$$

where I is intake of feed dry matter (DM) or a component such as organic matter (OM), nitrogen (N), etc., and F is the corresponding output in faeces.

Digestibility is a measure of the greatest and most variable single excretory loss to the animal of the energy and nutrients in the feed consumed. The maximum value of D for temperate grassland herbage is about 85 percent, and the minimum for cultivated species is rarely less than 45 to 50 percent. Corresponding values for tropical materials are about 80 and 30 percent respectively; Minson and McLeod (1970) found that, on average, digestibilities were less than those for temperate species by about 13 percentage units. The major cause of variation within these ranges is stage of maturity of the plant. There are differences between plant species, and legumes are generally more digestible than the grasses. Effects of fertilizers are principally indirect, resulting from changes these may induce in the botanical composition of the pasture, but when plants given no fertilizer have low concentrations of N or of mineral nutrients required by animals, additional supplies may increase digestibility and intake by the animals. Digestibility varies little with changing physiological state of the animal (e.g. age, pregnancy) but does decrease as the level of feed intake increases; Blaxter (1961) reported :

$$\triangle D = 0 \cdot 119\,(100\text{-}D_m) \qquad (2)$$

where $\triangle D$ is the depression in energy digestibility for a feed when intake increases from the maintenance level to twice maintenance, and D_m is the digestibility at maintenance. Later studies (Van Soest, 1975) indicate that changes in D with level of feeding are related closely to changes in the digestibility of plant cell wall substances and that a more generally applicable relationship than equation (2) can be based on analyses of feeds for these constituents.

Small and variable differences between sheep and cattle in digestive ability have been reported, probably reflecting the difficulties in establishing equivalent levels of feeding, but there is evidence (Blaxter, Wainman & Davidson, 1966) that cattle are superior in digesting roughages.

Equation (1) may be written :

$$I = F \times \left\{ \frac{100}{100\text{-}D} \right\} \qquad (3)$$

which is the form used for estimating grazing intake of usually DM or OM from determinations of F and D. Alternatively :

$$I = F \times A \qquad (4)$$

where A, the feed : faeces ration and termed the Intake Factor, is the reciprocal of the *in*digestibility as a decimal fraction. It follows that the intake (I_d) of digestible dry or organic matter (DDM, DOM) is given by :

$$I_d = F \times (A - 1) \qquad (5)$$

2. *Gross energy*

The gross energy (heat of combustion) of most grazed feeds reflects that of carbohydrate material (about 17·2 kJ/g), the major component. Values for herbage from mixed species pastures in New Zealand varied from 18·1 to 19·1, with a mean of 18·7 kJ/g DM (Hutton, 1961). Rather higher values in the range 18·2 to 20·1, with mean of 19·1 kJ/g DM, were reported for various pasture species in Britain (Armstrong, 1964) and Tasmania (Michell, 1974). Wider ranges of 17·3 to 20·5 (mean 18·7) kJ/g DM have been reported by Butterworth (1964) for forages grown in the tropics and of 17·2 to 18·7 kJ/g DM for two subtropical grasses and a legume by Minson and Milford (1966). Some of the variation arises from differences in ash content. Thus a forage with a gross energy of 18·0 kJ/g DM and containing 90 percent OM will have a gross energy of 20·0 kJ/g OM, and so will one of 19·0 kJ/g DM containing 95 percent OM. Variation is also positively related to content of crude protein, which has an average heat of combustion of about 23·6 kJ/g.

Some plants such as sagebush (*Artemisia* spp.) have a high content of essential oils and so high gross and digestible energy values, but the energy available to the animal (metabolizable energy) is lower because these substances are excreted in the urine (Cook, Stoddart & Harris, 1952).

3. *Digestible energy*

Hutton (1962) reported energy values for the DOM of temperate herbages in the range 18·2 to 19·5 and Michell (1974) values of 18·8 to 21·6 kJ/g. Most of the values reported by Minson and Milford (1966) were in the range 17·6 to 19·2 kJ/g, though some individual values were up to 8 percent less or 4 percent greater than this range; energy values for DDM were about 5 percent lower and were less variable. As with gross energy, this variation mainly reflects differences in the crude protein content of the digested feed.

There is an almost 1:1 relationship between percent DM digestibility (X) and percent energy digestibility (Y). Moir (1961) reported :

$$Y = 1{\cdot}006\,X - 2{\cdot}013 \quad (\text{RSD} \pm 0{\cdot}54)^{\star} \qquad (6)$$

Minson and Milford (1966) reported rather similar and equally precise equations for their subtropical forages, though significant differences between the regressions indicated that values calculated for these feeds from the general equation (6) may be biased. The maximum discrepancy between predicted Y values over a range in X of 50 to 70 percent, was 2·9 units. Discrepancies were smaller when predictions for the three subtropical forages were derived from OM digestibility.

The digestible energy content of forages (DE, kJ/g DM) can be estimated from DM digestibility using Moir's (1961) equation :

$$DE = 0{\cdot}193\,X - 0.661 \quad (\text{RSD} \pm 0{\cdot}26) \qquad (7)$$

Heaney and Pigden (1963) and Michell (1974) reported similar relationships. Values predicted from equation (7) and those for the three forages examined by Minson and Milford (1966) were in good agreement at values for X of about 55 percent or less, but at higher digestibilities the DE predicted by equation (7) were about 7 percent greater; an equation of similar form for some tropical forages (Butterworth, 1964) when applied to subtropical feeds overestimated DE by about 10 percent.

⋆ Residual standard deviation (or standard error of estimate).

4. *Metabolizable energy*

Metabolizable energy (ME) of a feed is defined as gross energy minus energy of faeces, urine (U) and gaseous products of digestion (principally methane, CH_4). Thus :

$$ME = DE - (U + CH_4) \quad (8)$$

Blaxter (1964) stated that " the metabolizable energy of grass and grassland products can be estimated with little error by multiplying the apparently digested energy . . . by 0·82 ". The factor 0·82 indicates that 18 percent of DE is lost to the animal in urine and CH_4; these losses, however, are variable. Equations have been published for estimating urine and methane as percentages of feed gross energy from, respectively, feed N concentration (Blaxter, Clapperton & Martin, 1966) and feed digestibility and level of feeding (Blaxter & Clapperton, 1965), and for relating ME to OM intake and digestibility (Graham, 1969).

Probably the most generally useful means currently available for calculating the ME content (kJ) of roughage feeds is from the expression [(14·23×g DOM) +(5·86×g digestible crude protein)] or where there is no information on DCP, then ME=15·06×g DOM (van der Honing, Steg & van Es, 1977).

5. *Energy retention and net energy*

Energy retention (ER; also called energy balance) by the animal is :

$$ER = ME - H \quad (9)$$

when H is the heat production of the animal (heat of cellular metabolism plus heat of fermentation in the digestive tract). Values of ER are positive and negative for animals on feed intakes respectively above and below maintenance levels. The net energy value of a feed is the increase in energy retention resulting from a defined increase in the intake of that feed.

At maintenance when net energy retention by the tissues of the animal as a whole is zero :

$$ME = H \quad (10)$$

Hence, if grazing animals with approximately constant liveweights are at maintenance, estimates of DOM intakes that can be made by techniques described below (expressed as ME = DOMI g/d × energy value) and of heat production (kJ/d) should be similar. ME intakes in excess of maintenance are converted to body energy gains and heat; the efficiency of conversion is variable, but with grazed feeds may be around 40 percent for body weight gain and about 60 percent if the energy is used for milk production, i.e., heat production is respectively about 60 and 40 percent of the ME.

Examples of the use of this type of information are to be found in the assessment in terms of energy flows of results from a grazing experiment (Hutchinson, 1971), and in a computer model of a pastoral production system (Vickery & Hedges, 1972a, b).

6. *Digestible protein*

When the crude protein content (X) of the feed is known, percent digestibility (Y) may be estimated by the equation of Glover, Duthie and French (1957) :

$$Y = 70\,(\log_{10} X) - 15 \tag{11}$$

This equation overestimated Y for subtropical forages of low protein content studied by Milford and Haydock (1965) who reported :

$$Y = 68{\cdot}03 - 284{\cdot}9\,\exp(-0{\cdot}3604\,X) \tag{12}$$

The percentage apparently digestible protein in a feed (Y_1) may be calculated from the protein content by the equation of Holter and Reid (1959) :

$$Y_1 = 0{\cdot}929\,X - 3{\cdot}48 \quad (\text{RSD} \pm 0{\cdot}46) \tag{13}$$

This equation appears to be satisfactory for both temperate and tropical forages (Milford & Minson, 1965).

7. *Chemical indices*

There are innumerable equations relating chemical component(s) of a feed to its digestibility or to other measures of nutritive value. In general, each of these is only applicable within a very restricted range of feed types, and even then the precision is low. For example, the digestibility and N content of herbage are negatively related, but Raymond (1959) reported that in 260 digestibility studies at one location, various feeds containing an N concentration of 2·5 percent ranged in OM digestibility from about 53 to 83 percent; he commented that one could easily have done better by guesswork.

Equations based on chemical analyses that differentiate plant cell contents, almost wholly digestible, from cell wall constituents (cellulose and lignin, etc.) which are only partially digestible or are totally indigestible, show more promise than those based on the Weende scheme of analysis (crude protein, ether extract, crude fibre, ash and nitrogen-free extract). Van Soest (1967) and Van Soest and McQueen (1973) reviewed the newer methods and their application for the evaluation of feed quality. Rather similar methods are described by Gaillard (1966) and Gaillard and Nijkamp (1968); Osbourn *et al.* (1971) indicated areas for improvements.

Proposals have been made for estimating the crude protein content of the feed from the nitrogen (N) concentration in the faeces, but this procedure is subject to the large errors associated with the prediction of DM and OM digestibility from faecal N, which is discussed later in this chapter under Faecal Index Techniques. Winks and Laing (1972) have used the concentration of crude protein ($N \times 6{\cdot}25$) in faeces DM as a criterion for supplementary feeding of cattle with urea mixed with molasses; no responses in liveweight to the supplement occurred unless faecal protein was less than 8 percent.

IV. ANIMAL MANAGEMENT

Animals used in studies of the productivity of vegetation should not be regarded just as defoliators. Competent and effective use depends as much on a sound knowledge of animal husbandry as on scientific expertise.

Even with animals that are uniform with respect to age, liveweight, previous treatment and physiological state, there will be considerable variability in performance. Variability in grazing intake and in production as a coefficient of variation, is unlikely to be less than 10 percent and values often of about 20 percent can be expected, though for digestibility values obtained by conventional means at a standard level of feeding, the CV is usually only 1 to 2 percent. Measurements of production should therefore be made with groups of animals that are as large as is practicable. Some variability can be minimized, e.g. animals with clinical signs of disease will obviously be excluded or treated before inclusion, though parasitism of the alimentary tract is normal. Unless the objective is, for example, to study the effect of some grazing management practice on parasitism, anthelmintics should be administered routinely and frequently, and it is advantageous to vary the type used from time to time. Similarly, minerals such as cobalt or selenium should be administered in areas where deficiencies are known or suspected, again provided that these are not the object of study. Other precautions should be taken to maintain health, such as vaccination against clostridial diseases, for the intrinsic worth of an animal in an experiment increases as the work progresses. Physical examination should also be made; obviously a study of sheep lactation should not include a ewe with an udder that has been injured during shearing.

Losses will occur, and replacement animals should be kept in conditions similar to those of the experiment. This is an important precaution because of the major effects of stocking density on the results obtained for production per unit area and per animal (Chapter 6).

Occasionally, an animal will be found unusually fractious and uncooperative, perhaps breaking fences or persistently damaging equipment it has to carry; such an animal should be excluded. A charge of biased selection might be justified if a large proportion of animals was excluded because of their behaviour, and in that instance the experimenter may well be at fault. Bad handling is wrong both for humane and for experimental reasons. It must not be assumed that animals are inherently uncooperative and have to be restrained by brute force in unnatural, uncomfortable positions for measurement or sampling. Most will respond to sensible handling and will soon come to accept a surprisingly large range of procedures; these should be standardized so that the animals become accustomed to a routine. Reports such as those of Reid and Mills (1962), Graham (1962), and Gartner *et al.* (1969, 1970) show significant metabolic responses in animals to unfamiliar environments and management, reflected for example in higher heat production and changes in the concentrations of many metabolites and other constituents in blood.

Training will minimize such effects, but it should be remembered that animal performance may still be affected by the physical interference and disturbance of normal behavioural patterns entailed in experimental routines. The latter will occur, for example, if animals that are normally gregarious are grazed individually on separate areas : Southcott, Roe and Turner (1962) showed that the productivity of a flock of only two Merino sheep was less than that of flocks comprising four or more. Observations on animals that do not have gadgets for the collection of samples will provide an indication of effects of physical interference, though not an absolute measure; these controls will inevitably suffer some disturbance if they are grazed with the experimental animals, but at the same time will not provide a truly valid comparison if grazing separately.

In addition to training the animals before the application of the experimental treatments, measurements should be made under uniform management conditions. These will provide information that can be used for stratified random allocations to treatments and data for subsequent use in covariance analyses. A preliminary period of uniform treatment will also reduce variation in animal performance during the main experiment that might otherwise have resulted from differences in previous management. For example, it might not be realized that some of the animals obtained at the start of a grazing season had been kept on a restricted diet and that rapid liveweight gains on a pasture, compared with others in a group that had been well fed, merely reflected compensatory growth after this previous management rather than any differences in pasture productivity. Allden (1968) reported that a mean difference of 8·1 kg in liveweight between two groups of sheep 1 year old, following differential nutrition, was rapidly reduced on a good pasture where the lighter animals consumed significantly more herbage. Such effects would be expecially misleading in short-term studies of pasture productivity.

As experimental subjects, mature castrates are relatively easy to handle. They have a rather constant feed requirement and temporary feed shortages present fewer problems than, for example, with ewes in late pregnancy; also, it is relatively easy to collect faeces free from urine for estimations of intake. The class of livestock that should be used depends on the type of study. For many pastoral situations it may be most appropriate to employ breeding females so that measurements are made of reproductive performances, and not only of liveweight gains and associated variables.

Each animal should be permanently and clearly marked for easy identification. This can be done by attaching ear tags, neck collars, hessian, etc., carrying numbers, symbols or colour codes. The animal itself can be marked with paint or, in the case of sheep, a substance that can be removed during wool-processing. Fire- or freeze-branding (Day, Jenkinson & Walker-Love, 1971; Pond &

Pearson, 1971) is often used on cattle.

The physical facilities available must allow easy and expeditious handling of the animals and minimize risk of injury to them or the operators. It is a false economy to build structures that are flimsy and badly designed so that animals can be moved only with difficulty, and possibly danger also, to areas where measurements and samples can be taken safely. Much time and effort is saved by sensible designs of yards (Plate 14), races, crushes and scales, arranged so that animals will naturally ' flow ' through them and can be restrained gently as necessary. Some shelter should be provided for the operators; water, and if possible electric power, should be available, and a complete kit of essential equipment (e.g. scissors, plastic bags, spares for ' gadgets ') and stock medicines used routinely (e.g. anthelmintics) or for such contingencies as blow-fly strike on sheep, should be kept on site.

Additional details on matters discussed in this section are given in publications by the Grassland Research Institute, Hurley (1961), Moule (1965), and Alexander (1973); the husbandry and handling of wild animals is discussed in Golley and Buechner (1968), and of deer by Drew and Kelly (1975).

PLATE 14. Easily erected yards for sorting cattle into groups. The steel panels can be moved to modify the size and shape of pens

V. PRODUCTION

Several measurements of production are often made routinely, including the yield and composition of milk of dairy cows, fleece weights of sheep at shearing, and liveweight changes. Some aspects of these measurements and others that are less easily made are discussed in this section.

1. Liveweight and body composition

Although liveweight is one of the most readily obtained and informative measures of animal performance, its measurement is problematical. Errors and possible bias in estimates of gains can readily arise from fluctuations in the quantities of digesta in the alimentary tract; for example, day-to-day and within-day changes of several kg in gut fill and hence in liveweight can occur with adult cattle owing to variation in the quantity and quality of feed eaten and the amounts of water drunk. Variation in the interval between time of weighing and the last defaecation or urination is of relatively minor importance; this effect tends to assume a constant pattern within groups of animals because they often adopt the habit of excreting when assembled for weighing which should, however, be done as rapidly as possible.

Large changes in gut fill often follow a change in diet, as from winter rations to spring pasture; a move from one pasture to another, say from grassy to leguminous herbage, could have a similar though perhaps smaller effect. Moreover, cattle or sheep kept on these two pasture types, though similar in liveweight, may differ considerably in gut fill. Treatment groups should be similar in mean age as well as in initial liveweight because fill varies with age ; it is small (<5 percent) in young animals and increases as solid feed intake increases and the rumen develops, but tends to decrease in adults as they fatten.

The quantity of feed eaten, reflecting the amount available on the pasture, and the time it was eaten in relation to the time of weighing, have major effects on fill and observed liveweight. For example, in a rotational grazing system with moves to a fresh pasture area each week, fill could be expected to be greater at the beginning than the end of each period; mid-week weighing should therefore reduce variability. Similarly, within-day variation could be expected in a strip grazing system with daily shifts of an electric fence.

There are several ways of standardizing weighing procedures so that errors and biases are minimized (Grassland Research Institute, Hurley, 1961; Moule, 1965; Campbell, 1969; Alexander, 1973). This subject has been reviewed comprehensively by Hughes (1976). At the end of an experiment, animals from several treatments could be grazed together for a week or so to promote uniformity in fill, though at the risk of diminishing treatment effects. In any event, the time of day chosen for all weighings should be related to the pattern of grazing behaviour of the animals. Hughes and Harker (1950) and Taylor (1954) in England found that day-to-day variation in cattle weights was least if these were recorded 3 to 4 h after sunrise, which is usually the start of a period of intensive grazing, the actual time of day of course varying with season of year. Indeed, the level of variation is then comparable with that obtained by fasting overnight. This would not be so if the water load in the pelage was increased by rain, or if adverse weather disturbed the normal behaviour pattern; weighing could then be deferred to another day. Several studies have shown that there is little or no gain in precision from averaging liveweights recorded on three successive days, a procedure that increases disturbance of the animals and labour requirement. It is better to increase the number of animals than the number of weighings. Additional precision is gained by fasting the animals overnight, but drinking water need not be withheld (Hughes, 1976). This practice is not always desirable because effects on the animals' performance may persist for some time after a fast, but it is a useful preliminary to determining start and finish liveweights which are particularly important in short-term trials. It is especially useful if the various grazing treatments are likely to have caused differences in fill and it is not possible to slaughter the animals and assess responses in terms of carcass gain.

Although liveweights have been determined by best procedures, further problems arise because the chemical composition of liveweight gain or loss varies. At least a three-fold variation is possible in the energy value between a gain made at low liveweights by young or lean animals and the same gain made by heavy, fat animals. The animal at constant liveweight may not be in the state of maintenance (eqn. 10) and if, as has been observed by Blaxter, Clapperton and Wainman (1966), nitrogen is retained though fat is being mobilized, there may be a substantial shift in body energy content. This is because the heats of combustion of fat and protein tissues are about 40 and 4 kJ/g, respectively, the latter containing about four parts water to one part protein.

Methods have been developed for estimating the chemical composition of the live animal. These are based on the observation that the fat-free empty body of animals of a given species tends towards constancy in composition with respect to water, protein, and mineral matter (ash) (Burton & Reid, 1969; Searle & Graham, 1972). Since relatively little water is retained in the fat deposition, increases in the amounts of fat in the body are reflected in a decreased water content, and so an estimate of total body water (TBW) in an animal of known liveweight is an index of its composition in terms of fat, protein, ash and energy value. TBW as estimated in the live animal will generally include water in the gut contents; variation in this is minimized by fasting without water for 24 to 48 h, or for shorter periods especially with young or undernourished animals. A marker substance that becomes distributed throughout the body water is then injected and after allowing time for equilibrium in distribution, which varies with size of animal and route of administration, the concentration of the injected substance in the water extracted from a blood sample or a series of samples is measured. The value that would have been obtained at the time of injection (t_0) had mixing been instantaneous is determined from these results. TBW (ml) is then calculated by dividing the amount of marker injected by the quantity present per ml of the water at t_0.

The substance now most commonly used is tritiated water (TOH)*. Equilibration time for sheep is about 6 h and the specific radioactivity (i.e. concentration) then is often taken to be the t_0 value because losses of this marker in urine, sweat and from the lungs will have been small unless ambient temperatures are high; it is generally advisable to take a second blood sample about 10 min later as a check on the first result. The stable (non-radioactive) nuclide deuterium (as D_2O) may also be used, but assay is considerably more time-consuming. TBW estimated from TOH (TOH space) is usually greater than the value determined by desiccation, but is most simply used without adjustment as one independent variable, with fasted liveweight as a second variable, in equations for predicting body composition. This has been done, for example, by Searle (1970) in a study of the chemical composition of sheep from three days to 18 months old. When published equations are used, care should be taken to conform with the conditions of derivation, including length of fast and time(s) of sampling, and to ensure that the dose injected is prepared exactly and that the whole dose is indeed injected.

Residual standard deviations from the equations of Searle (1970) for predicting fat, protein and ash were $\pm 0 \cdot 64$, $0 \cdot 24$ and $0 \cdot 15$ kg respectively, and for energy was $\pm 19 \cdot 1$ MJ, which is equivalent to 1 to 2 kg body gain in older sheep. Donnelly and Freer (1974) found that more generally applicable equations having greater precision were obtained by including an index of mature body size as an independent variable. Reardon (1969) calculated that an estimate of energy retention in grazing sheep derived from TOH measurements at the start and end of an experimental period would have a standard error of about $6 \cdot 5$ MJ; that obtained by the comparative slaughter technique (slaughter initially of a sample of animals, and of all at the finish) would have an SE more than twice as great. However, the most precise estimate would probably be obtained by combining the two methods, TOH plus sample slaughter at the start and

---

* Tritium (^3H) emits β particles of very low energy and is considered among the least toxic of all radionuclides. Its half-life is 12·6 yrs, but the rate of clearance of TOH from the animal is rapid so that the *biological* half life is only one to two weeks. Accumulation of TOH in field environments is unlikely even after repeated dosing of animals because of dilution with rainfall and by exchanges of water between soil, plant and atmosphere. Similar considerations apply with other radionuclides commonly used in biological research. Government regulations usually prohibit sale for human consumption of animals that have been dosed with radionuclides. In any event, permission must be obtained from the appropriate authority before these are used in the field or elsewhere.

slaughter at the end. The TOH or similar techniques may be inappropriate during the course of certain experiments owing to the effects of the fasting required.

Up to the present time there are few equations for cattle (Little & Morris, 1972; Crabtree, Houseman & Kay, 1974) owing to the cost and practical difficulties of determining chemical composition.

Measurements such as the linear distances between skeletal reference points, and assessments of body condition can provide useful information (Grassland Research Institute, Hurley, 1961; Moule, 1965; Alexander, 1973). Liveweight alone may be insufficient to indicate when an animal is in a suitable condition for slaughter and might be removed from the experiment, and would not in itself distinguish between a large-framed but thin animal and one that was small and fat and had substantial body reserves of energy. Height at withers, length of cannon-bone, and length of body (e.g. withers to pin-bone) help to describe conformation, the relative proportions of the parts of the body; heart girth, chest depth, and widths at shoulders, ribs and hips help to assess fleshing or ' finish '. Methods of making these measurements on cattle and their accuracy are described by Fisher (1975), but they are of little value for predicting carcass composition (Ruohomäki, 1975; Berg & Butterfield, 1976). Subjective assessments of body condition of sheep and cattle can be informative, provided descriptions of grades are rather explicit. Russel, Doney and Gunn (1969) palpated the lumbar vertebral region of sheep to appraise the prominence and degree of cover of the spinous and transverse processes, the depth of the *Mm. longissimus dorsi* and the subcutaneous fat cover. The condition scores, as defined in terms of grades in the range 0 = extremely emaciated to 5 = processes cannot be felt and very thick fat, etc., were closely related to the quantities of fat in the body as determined by chemical analysis after slaughter.

When an animal is slaughtered at the end of an experiment, estimates can be made of the composition of the carcass without its total destruction. Density, determined by weighing in air and in water, may be used as predictor (Lofgreen, 1965) though care must be taken to standardize the conditions under which the carcass is chilled and the temperature of the carcass and the water in which it is weighed. For example, an error of only 0·001 in density alters by about 1 percent the estimated energy content of the empty body as predicted by an equation of Garrett and Hinman (1969). Large errors can occur if analyses of sample joints such as the 9-10-11 rib cut are used to predict body composition, though these may give useful information on proportions of muscle, fat and bone in the carcass (Stouffer, 1969; Preston & Willis, 1970; Ledger, Gilliver & Robb, 1973). The CSIRO Division of Tropical Agronomy, Brisbane, uses as a quick estimate of the ' finish ' of cattle the thickness of the fat over the eye muscle at the 12th rib (Plate 15); this measurement is an indication of overall carcass composition (Charles, 1964; Charles, Butterfield & Francis, 1965). In the Australian beef carcass appraisal system (Johnson & Charles, 1976), percentage of fat is estimated from carcass weight and the thickness of subcutaneous fat over the cut surface of *Mm. longissimus thoracis et lumborum* between the 10th and 11th ribs. An adjustment is made for carcass weight so that for a given increase in fat thickness the estimate of fat percentage is higher in light-weight carcasses. Merit for muscle is based on the approximate length and the cross-sectional area of *M. longissimus* (rib-eye area) between the 10th and 11th ribs, relative to the weight of the fat-free carcass. Scores increase with increasing carcass length and rib-eye area up to specified measurements for prescribed fat-free carcass weights.

2. Milk

Milk is a sensitive indicator of the nutritive value of the diet, and changes in the diet are quickly reflected in changes in milk yield and composition. For example Corbett and Boyne (1958) and Jeffery (1970) found that when concentrates were given to grazing dairy cows, the responses in milk production were essentially complete within four days. A reduction in dietary energy intake soon results in decreased milk yield, followed by decreases in protein and fat content; there may also be a change in the proportion of the component fatty acids, $C_{4\text{-}16}$ acids decreasing and C_{18} increasing (Storry, 1970; Stobbs & Brett, 1972).

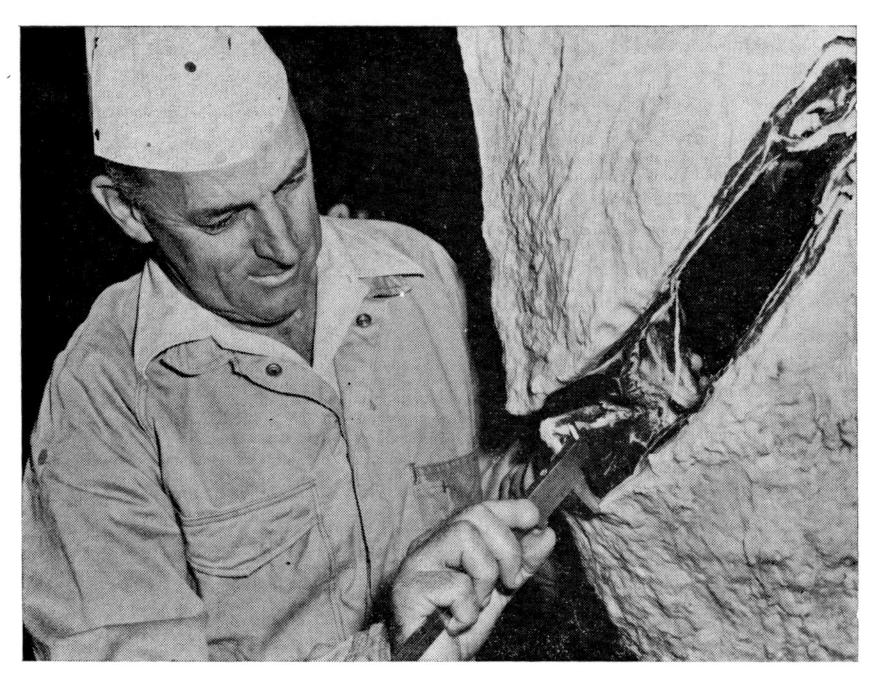

PLATE 15. Measuring fat cover over the eye-muscle at the 12th rib using a vernier caliper

The energy value of milk (Y, kJ/kg) can be estimated as follows :
for cows (Tyrell & Reid, 1965)

$$Y = 386\,F + 205\,SNF - 236 \qquad (RSD \pm 37) \qquad (14)$$

for ewes (Brett, Corbett & Inskip, 1972)

$$Y = 328F + 2{\cdot}5D + 2203 \qquad (RSD \pm 140) \qquad (15)$$

where F and SNF are the fat and solids-not-fat percentages (w/w), and D is the number of days after lambing.

To determine the milk yield of beef cows (Alexander, 1973) and sheep or other animals (Golley & Buechner, 1968) and obtain a representative sample for analysis may present some difficulty. Yield may be estimated from the weight gains of the young during periods of suckling, but composition determined on the small samples that can be obtained during these periods is likely to be unreliable. The young are prevented from sucking between weighing periods by separating them from their dams, or by covering the udder (Owen, 1957), or by fixing a rectangular metal flap to the nose as done by Bluntzer and Sims (1976) with calves. It is important that the number of weighings made in a day should be similar to the number of sucklings in normal behaviour, because if the number is less the milk yield may be reduced (Wheelock & Dodd, 1969). Walker (1962) observed that beef calves were suckled three to five times in 24 h, but the frequency changes with age (Nicol & Sharafeldin, 1975) and may be greater with sheep (Ewbank, 1967). Labour requirement for the ' suckling ' method is high, because each set of recordings made at intervals during the course of lactation should extend over 24 h. There are some other problems. Lambs and calves may defaecate or urinate at feeding times and, apart from this, as the probably lively animals grow it becomes increasingly difficult to determine accurately liveweight differences as little as perhaps only 1 percent between the start and end of a suckling period. It has also to be assumed that the frequent disturbance entailed in making these measurements causes no change in performances of either dam or young at or between sucklings.

Gifford (1953) estimated the milk production of beef cows during a period of three days in each month when their calves were allowed to suck twice daily. On the second day, one half of the udder was milked while the calf sucked the other half, and on the following day the opposite half was milked. The milk obtained was weighed and analysed, and the sum of the two ' half ' estimates was taken to be one day's production.

Yield can be determined by milking the whole udder while the young and dam are separated for some hours. With animals unaccustomed to this procedure it is often necessary to obtain let-down of the milk by intravenous injection of oxytocin (posterior pituitary extract). For this purpose a single dose of 3 I.U. has been found satisfactory for sheep (Corbett, 1968) and 20 I.U. for cattle (Lamond, Holmes & Haydock, 1959). A determination requires two milkings, the first to remove the variable amount of milk in the udder and the second about 4 h (sheep) or 6 h (cattle) later, when the milk that has been secreted during this interval without suckling is obtained. The interval in minutes, most easily defined by the times of oxytocin injections, is recorded, as also is the quantity of milk obtained at the second milking. It may be easier and no less accurate in the field to measure yield by volume and convert to weight by a density factor (1·035 g/ml) instead of weighing directly. The milk should be thoroughly mixed to obtain a representative sample for analysis.

It is important that at both milkings the udder is emptied to a standard degree. With ewes this is probably achieved more surely, and certainly more easily and quickly, by machine- rather than hand-milking, but with cows Lamond, Holmes and Haydock (1969) recommended cannulation of the teats. One injection of oxytocin at the start of milking suffices; let-down indicating competent injection, is obvious within 30 sec and milking of a ewe is complete at about 1 min and of a cow at about 1 to 4 min later. A second injection during the same milking is not recommended because this will allow removal of some of the residual milk normally always present in the udder, with subsequent effects on yield and composition (Wheelock & Dodd, 1969). Teat cups suitable for ewes are obtainable from milking machine manufacturers (see Treacher, 1970); the pulsation rate should be considerably higher than for cows, in the range 120 to 170 per min and the vacuum lower (28 to 34 cm mercury).

No significant diurnal variation was found in either the yield or composition of milk of grazing ewes (Corbett, 1968) or the yield of grazing cows (Lamond, Holmes & Haydock, 1969), so that 24-h production can be simply calculated by multiplication from the 4- or 6-h values. This does not hold for housed animals given feed intermittently, under which conditions considerable diurnal variation is observed (Ryley & Gartner, 1962; Robinson, Foster & Forbes, 1968).

The milking technique measures the rate of milk secretion and not necessarily the amount consumed by the young. Consumption may be much less than production in early lactation if the dam's milk production potential is high. Moore (1967) observed that another cause of differences between estimates of production and consumption in sheep was ewe behaviour. With ewes that stood quietly when the lamb was allowed access, the recorded intake was similar to the estimated production of the same animals, but with ewes that were restless and did not allow uninterrupted suckling by lambs until these had obviously finished, yields by milking were the greater.

The general shape of the lactation curve and lactation yield within the particular nutritional regime, and the rates of secretion observed after early lactation when the curve is intermittently monitored, will be determined mainly by the mutual behaviour of dam and young during the time these are left together undisturbed. Milking only once or twice weekly, i.e. involving disturbance for only 4 or 8 h, can provide adequate information.

There is a significant positive correlation between milk production and growth of the offspring while these are still eating little or no solid feed (Wallace, 1948; Owen, 1957; Peart, 1968). Regression relationships have been established to predict production from lamb or calf liveweight gains, as done by Robinson, Foster and Forbes (1969) for housed sheep in a particular system of management, but at pasture such equations would be of little value later than 3 to 4 weeks post-partum. These may be useful during the early stages of lactation when the milking technique is least reliable, as may be another indirect method proposed by Macfarlane, Howard and Siebert (1969) which involves measurement of water turnover rate in suckled young. Milk contains about 85 percent water, and oxidation of the hydrogen in the solids yields metabolic water, so that total water yield is approximately 95 percent of milk volume. Turnover is measured by an adaptation of the TOH technique for estimating body composition. The young animal is injected with TOH and is returned to its mother after an interval of 2 to 6 h to allow for equilibration; a sample of blood is then withdrawn for radioactivity assay. Comparison of the result of this assay with one made 7 to 14 days later provides an estimate of the water turnover rate, which was found to agree closely with actual milk intake given by bottle in a controlled experiment.

In this TOH method, used in the field by Yates, Macfarlane and Ellis (1971) on calves up to 8 weeks old, it is assumed that intake of other feed and drinking water is negligible. However, grazing does commence within a few weeks of birth and will cause progressive over-estimation of milk intake. For example, reports of Langlands (1972, 1973) indicate that while grass intake by unweaned lambs is low at 3 weeks of age, it may account for half of total OM intake at 8 weeks. This problem has been tackled by Holleman, White and Luick (1975) who injected both dam and offspring, one with TOH and the other with D_2O. The method allows calculation of the percent water in the calf derived from the cow and then, by reference to the calf's body water pool and turnover rate, its milk intake can be calculated; the method has been used on reindeer that were grazing freely on natural pastures. Another double-marker method, but for shorter-term estimates, has been described by Nicol and Irvine (1973).The calf receives orally a known dose of ^{125}I, and from the cow previously injected with ^{131}I it ingests milk of known radioactivity; these data, with measurements of the subsequent $^{125}I/^{131}I$ ratio in the calf's blood, allow calculation of milk intake.

3. Wool

Wool growth is positively related to feed intake and continues, though slowly, even when intake is below the maintenance level. Rate of growth shows seasonal variation (photoperiodic, temperature) to a degree varying with sheep genotype; for example, it is smaller in the Merino than in Scottish Blackface or Cheviot breeds (Doney, 1966), but there is also important variation due to seasonal changes in herbage growth and to type and management of the pasture (e.g. Langlands & Bennett, 1973b; Langlands & Bowles, 1974; Patil, Jones & Hughes,

1969). Fleece weights at shearing usually give production over a year, but production over shorter periods is often of interest and will allow correction of observed changes in liveweights to a standard shorn state.

There are two main methods of measuring wool growth over successive periods of about 3 weeks minimum length. In the first, wool is harvested by small animal clippers from an area, usually 10 $\times$ 10 cm, delineated by tattooed lines on the midside (Plate 16) (Moule, 1965). In the second, a dye solution is applied on the skin after parting the fleece along a dorso-ventral line about 10 cm long on the midside; further applications are made at intervals at the same site. Full details are given by Chapman and Wheeler (1963) and Williams and Chapman (1966) for determining, from the distances between the resulting dye-bands on the staples, successive rates of growth in terms of greasy or cleaned scoured weight and, if required, fibre diameter values. The dyebands are sharply defined in Merino fleeces but may be unsatisfactorily diffuse and impermanent in the wool of other breeds. Wool that has formed in but has not emerged from the follicles is not dyed or harvested, and if growth is to be measured before and after particular events such as lambing, dyebanding or clipping should be delayed for some days. Downes and Sharry (1971) reported that emergence time varied from 5 to 10 days and could change by as much as 4 days with a change in feed intake. They also reported that growth in length on an individual sheep was closely related to the fibre diameter. Their studies employed a radioautographic technique involving intravenous injection of L[^{35}S] cystine which allowed measurements of wool growth to be made over periods as short as 2 to 3 days.

Measurements over longer periods will usually be adequate. Langlands and Wheeler (1968) found that the dyebanding and tattooed patch techniques appeared to be of similar precision, and that the errors associated with them were small compared with the variation between sheep. They made the recommendation, which applies to studies on other forms of animal production, that wool growth should be measured during a pre-experimental period and used as a covariate. Of the two methods, dyebanding is the quicker and easier; sheep do not have to be prepared by tattooing, and clipping of a patch area can be time-consuming, especially if the skin is wrinkly. There is evidence (Wodzicka-Tomaszewska & Bingham, 1968; Downes & Hutchinson, 1969) that exposure and local cooling of the skin at the patch affects wool growth there. Dyebanded wool, which is permanently stained, must be carefully removed from fleeces sent for sale.

The fleece normally contains suint and wax in proportions that vary among breeds. Paladines *et al.* (1964b) reported that the heats of combustion of wool wax and wool protein (clean fibre) were 40·76 and 23·47 kJ/g respectively, and the percent clean wool in the fleece, termed ‘ yield ’, is determined by ‘ scouring ’ as described by Chapman (1960). It is particularly important to determine yield for wool from dusty environments where fleece weight can be increased greatly by extraneous mineral matter.

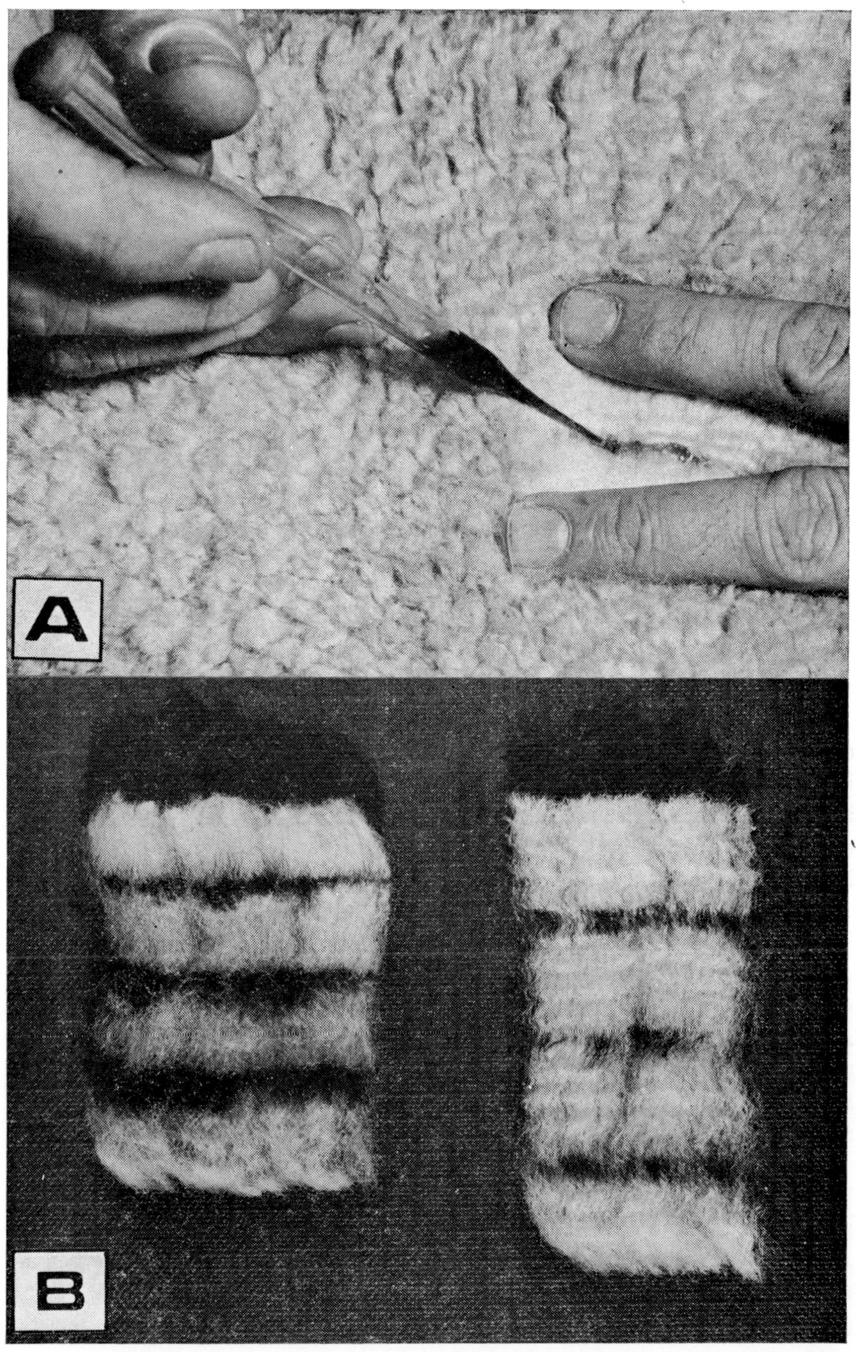

PLATE 16A, B. Measuring wool growth from dye-bands (Chapman & Wheeler, 1963) A. Dye solution being applied at skin level. B. Dye bands applied over intervals of time.

4. Reproduction

There is a variety of indices of reproductive performance, and care must be taken to define the bases explicitly. For example, the number of lambs born can be expressed as a percentage of the number of ewes joined with the rams, or of the number that actually mated, or of the number that lambed. Similarly, the ewe flock can be described in terms of the percentage of those joined with rams that mated, the percentage of those mated that conceived, and the percentage of those conceiving that lambed. Observations to be made on the young include percentage survival, birth weights, liveweight gains to weaning, the numbers weaned, and the total weight of weaners in relation to the number of ewes or cows, etc.

In long-term studies involving breeding animals, their productivity may alter owing to changes in the size and structure of the flock or herd. These changes are described by constructing age-specific tables (Moule, 1965; Golley & Buechner, 1968) that include death rates for the various age groups, and female fecundity rate, which is the number of female offspring born at each age. The sum of the age-specific fecundities constitutes the gross reproduction rate, which describes the average number of females (potential mothers) born to one mother during her reproductive lifetime, and the net production rate is the surviving female offspring produced during that lifetime.

A number of precautions will minimize the possibility that reproductive performances are affected by disease or physiological abnormality and do not truly represent the productivity of the area and animals. Veterinary examination of the male is discussed by Moule (1965), Foote (1969), Alexander (1973) and in Golley and Beuchner (1968). Though it may be established that the males to be used are healthy, produce satisfactory volumes of semen showing good motility and containing the normal density of normal spermatozoa, and that they have good libido and serving ability, they might suffer a temporary infertility while in use. They should therefore be interchanged among the groups of breeding females at intervals of one to two weeks during the period of joining.

Details of surgical and other techniques for diagnosis of pregnancy or reproductive disorders and wastage in female livestock are given by Moule (1965), Anderson (1969), Alexander (1973), and in Golley and Buechner (1968). Pregnancy diagnosis may be an important preliminary to establishing an experiment; for example, it is quite often found that females required to be and stated to be non-pregnant are in fact not so.

Oestrus can be observed in sheep by the marks from crayons attached to the brisket of the ram (Radford, Watson & Wood, 1960). and in other species by similar means, such as the chin-ball harness for bulls (Lang *et al.*, 1968; Lang & Hight, 1969) and the heat-mount detector on cows (Baker, 1965). These observations can be used to calculate embryonic loss: the number of females that mate and do not produce young is expressed as a percentage of all females that mated.

If oestrus is to be observed but pregnancy is not desired, vasectomized males or related techniques may be used (Golley & Buechner, 1968). First oestrus, signalling puberty, occurs when animals have grown to certain minimum liveweights rather than at particular ages (Dȳrmundsson, 1973a, b; Young, 1974) and so is an indicator of general nutritional level.

Hormonal control of reproduction is discussed by Owen (1974), Lamming, Hafs and Mans (1975) and Gordon (1975). Synchronization of oestrus can be used to promote uniformity in reproductive behaviour among experimental groups, simplifies the practice of artificial insemination, and shortens the time span of parturition. Another objective may be to obtain young animals of rather uniform age for a later experiment.

Changes in the level of nutrition of animals caused by variation in feed supplies may affect reproduction at all stages: age at puberty, spermatogenesis, ovulation, conception and implantation, pregnancy, parturition, and the number, viability and birthweight of the young. In addition, some plants contain substances that cause reproductive disorders, the most notable being the oestrogens present in certain cultivars of *Trifolium subterraneum* (subterranean clover) and in *T. pratense* (red clover) (Moule, Braden & Lamond, 1963; Braden & Baker, 1973). This should be born in mind when reproductive performances of grazing animals are poor, especially when there appear to be adequate supplies of feed and the animals are free of disease.

5. Livestock units

Many countries have developed systems for expressing the output from grasslands in terms of a standardized unit, such as grazing days, when interest lies in a single class of stock, or as ewe equivalents (New Zealand), dry sheep equivalents (Australia), large-stock units or 'standard' cows (Netherlands, etc.), or in terms of a calculated quantity of feed units, such as utilized starch equivalent (Europe) or total digestible nutrients (TDN) (N. America) (Barker *et al.*, 1955; Schneider, Soni & Ham, 1955; Coop, 1965; Campbell, 1969). It should be understood that the units are not, and are not intended to be, measures of actual herbage growth. They are an index of what, by inference from the types, numbers, liveweights and production of the animals on the pasture, has been harvested by the animals in grazing. The conversion of various animal outputs to a single common basis provides the grazier with a convenient guide to management and economic assessments. Comparisons on such a basis of outputs from the various pastures on a farm can indicate that one type of soil or pasture is less productive than another; or if outputs were less than from an adjacent farm of similar type, then some aspect(s) of management might be examined.

Generally, the units represent a reverse use of feeding standards : if the standards indicate that X kg of TDN are required by a cow for maintenance and the production of Y kg milk, than the production of Y kg by a similar cow at pasture represents the intake of a quantity of herbage that also contains X kg TDN. Equivalents among animals differing in types and size are assessed similarly. The results obtained are reasonably satisfactory for application in commercial practice on a comparative basis, provided that there is uniformity in the conventions adopted throughout the range of comparisons. For example, it is simplest to attribute to a supplementary feed its full theoretical productive value (e.g. 1 kg gives 2 *l* milk), and attribute only the residue of production and the maintenance of the animals to the intake of nutrients in grass. Again, in the scheme described by Barker *et al.* (1955), energy expended in grazing activity was ignored because it was considered that a uniform allowance would be inappropriate. Conventions such as these will lead to underestimation of herbage productivity under grazing.

During the past two or three decades there has been an increasing application of the techniques described in this chapter. There are now many reports on the grazing intakes (I) of animals and their corresponding liveweights (W), milk production (M) and liveweight gain (G), etc. Many of these reports have followed Wallace's (1956) use of multiple regression (or related) analysis, which was applied in the following formula by Brody (1945) to examine input/output relationships for housed dairy cows :

$$I = aW^{k} + bM + cG \qquad (16)$$

The values of the coefficients obtained by such a partitioning of the input amongst the outputs have been taken as measures of the animal's nutrient requirements; for example, the maintenance requirement is taken to be *a* kg of the intake (often expressed as kg digestible OM) per unit of liveweight (often $W^{0.75}$ or $W^{1.0}$). To what extent the coefficients have genuine biological meaning is open to argument (Corbett, 1969). On the other hand, provided there has been no important bias in the measurements of the variables, the equations can legitimately be used from ' right to left ' in situations where conditions of application are similar to those of derivation; that is, grazing intakes can be predicted from observations of W, M and G, etc. Many such equations have been derived from the results of studies on grazing cattle and sheep. Logan and Pigden (1969) listed a number, published in the early 1960s, that could be used to estimate the quantities of herbage grazed by several types of livestock in various pasture and environmental conditions. Conway (1973) examined the results of many published studies on the intakes under *ad libitum* grazing conditions of cattle varying in age, liveweight and lactation status, and derived a generalized equation relating maximum voluntary intake to liveweight and herbage DM digestibility. This equation was used to derive relative stocking values (units) for various cattle types.

In recent years, the development of computer modelling of grazing systems has given fresh impetus to the acquisition of objective information on climate/soil/plant/animal relationships. Synthesis of the models from a wide range of existing information and their continuing modification as new knowledge is gained, has set new dimensions for such concepts as livestock units, which can certainly be refined and made more objective and comprehensive. The units will continue to be used principally as a basis for management and economic decisions. When critical measurements or comparisons of pasture production and utilization are to be made, there must be direct investigations; these are, of course, required to improve existing models or build new ones, and for the crucially important validation studies associated with their development.

VI. NUTRITION

1. Faeces

Measurements of faecal output are required for estimation of grazing intakes (equation 3), and examination of faeces can provide information on gastro-intestinal parasitism and the botanical composition of the diet (see Chapter 3).

(a) Total collection

Methods for collecting faeces in bags attached to a harness worn by grazing animals are described by Sears and Goodall (1942), Ballinger and Dunlop (1946), Cook, Stoddart and Harris (1952), Lesperance and Bohman (1961), Arnold and Bush (1963), and Royal (1968) (Plate 17). This is a fairly simple procedure, but has some disadvantages :

PLATE 17. Equipment for collecting faeces (Ballinger & Dunlop, 1946)

(i) The animal's behaviour and performance may be affected, especially if the output of faeces is large and the harness is badly designed or adjusted so that there is pressure on or chafing of some part of the body. It is advisable to accustom the animals to the harness for some time before the actual collection period begins and to leave it on between these periods with the collection bag attached but left open at the bottom.

(ii) Collection from females is difficult owing to problems in separating urine. If DM only is to be measured, mesh bags made from a material such as woven Terylene can be used to collect from ewes provided their faeces are in fairly dry pellet form.

(iii) Measurements may be biased owing to incomplete collection which, even if known or suspected, cannot easily be quantified.

(iv) Cost of the harness and bags can be substantial, especially if for cattle; also the labour requirement for changing the bags, usually twice daily, and their maintenance and adjustment is quite high.

(v) By preventing the normal return of faeces to the sward, collection will tend to invalidate studies of pasture production.

(*b*) *Estimation*

If an animal is dosed regularly with X g of an inert reference substance or marker that is not digested or absorbed and is wholly excreted in the faeces, and the mean concentration in the faeces is found to be Y mg/g DM or OM, then :

$$F\ (g) = \frac{1000X}{Y} \tag{17}$$

Several substances have been used as markers, but chromium sesquioxide (Cr_2O_3) is the one most widely used. It will be seen from equation (17) that if a sheep is given 1 g Cr_2O_3 twice daily and the mean concentration in its faeces is 10 mg/g OM, daily faecal OM output is 200 g OM; with cattle given 2 $\times$ 10 g Cr_2O_3 daily, the same concentration indicates 2 kg faecal OM.

Unfortunately, the actual concentration varies considerably about the mean between times of day because the Cr_2O_3, given in discrete doses, does not become uniformly mixed with the contents of the digestive tract and the rate and pattern of its excretion differs from that of undigested feed residues. Biases in estimates of faecal outputs may be very large unless care is taken in the administration of the Cr_2O_3 and in the sampling of the faeces.

Administration, which should start not less than seven days before estimates are to be made, is most easily done by giving the dose in a gelatin capsule with a balling gun. Ready-made capsules containing 1 g Cr_2O_3 for sheep or 10 g Cr_2O_3 for cattle, are available commercially[1]. The gelatin is coloured orange so that capsules ejected by the animal and not swallowed can be readily seen on the ground: however, ejected capsules have usually been chewed so that the animal will have consumed an unknown proportion of the dose. This problem rarely occurs when a balling gun is properly used; nor, with proper use, will the gun cause injury to the animal. Unskilled workers should be carefully instructed and supervised for a period to ensure that the end of the gun is not thrust into the pharynx, resulting in painful injury, or possibly choking the animal by dosing into the trachea.

Once-daily dosing results in large within-day variation in faecal Cr_2O_3 concentration; twice-daily dosing at, say, 0900 and 1600 h is more usual. Variation is reduced if instead of gelatin capsules a special paper incorporating Cr_2O_3 is used (Corbett *et al.*, 1960). The paper[2] supplied in shredded form is less convenient because the individual doses have to be prepared by hand, but with a suitable balance can quite rapidly be weighed out with a tolerance of $< \pm 0 \cdot 1$ g ($0 \cdot 03$ g Cr_2O_3). The doses are wrapped in ordinary paper for administration.

The simplest sampling procedure is the so-called grab-sampling, which is manual removal of faeces from the rectum, and is done most conveniently at the time the animals are dosed. If there were no significant irregularities from day to day in the Cr_2O_3 dosing regime or in the animal's feeding patterns and behaviour, it might be possible to determine the time(s) of day when the Cr_2O_3 concentration in a grab sample happened to be the same as the mean daily concentration and thus to sample at these times. However, this procedure is not reliable with housed animals, let alone those at pasture. Lambourne and Reardon (1963a) dosed and grab-sampled grazing sheep twice daily, combining samples for individual sheep on an equal dry-weight basis. Their estimates of faecal DM outputs over periods varying from 5 to 14 days were similar to the actual outputs and had standard deviations of ± 12 to 14 percent relative to these. Using similar dosing and sampling routines, Langlands *et al.* (1963a) estimated the faecal outputs of cattle and sheep with similar precision, but showed that there were two important sources of bias. One is a discrepancy between grab and representative samples in Cr_2O_3 concentration; in the several experiments reported, the estimates of faecal outputs obtained by analyses of the two types of samples differed by $+5$ to -15 percent. The second source of bias is the failure of marker output to equal marker intake during the period of estimation; differences of $+2$ to -9 percent were observed between actual faecal outputs and those estimated from Cr_2O_3 in representative samples. Such discrepancies would

---

[1] R. P. Scherer & Co. Ltd., Bath Road, Slough, Bucks., England

[2] Enquiries to Dr. J. F. D. Greenhalgh, Rowett Research Institute, Bucksburn, Aberdeen AB2 9SB, UK.

occur if there had been faulty dosing, faulty collections or analytical errors. However, Raymond and Minson (1955) have argued that the Cr_2O_3 technique tends to estimate the quantity of faeces containing the amount of the marker dosed per 24 h, and that the quantity of faeces excreted during a period of a few days may not correspond to the marker dose if there are short-term changes in the rate of passage and period of retention of digesta. Compared with total collections, Cr_2O_3 estimates free from sampling bias could be regarded as more valid, particularly if the results are used for the calculation of feed intakes, because estimates are being made of the faeces derived from the intake over a standard period rather than of the faeces output as such over a standard period.

Estimation from Cr_2O_3 in grab-samples is the most widely used procedure, but it is advisable to establish whether there is bias, and if so what its direction and magnitude are. Bias is likely to vary from one experiment to another, and factors to correct for this can be obtained by harnessing a few animals and collecting as well as grab-sampling their faeces; Cr_2O_3 concentrations in representative and grab-samples are compared. Alternatively, if the collections are demonstrably complete and there is no reason to suppose irregularity in rate of faeces excretion, or if the collection periods are sufficiently long to minimize effects of any such irregularity, actual outputs can be compared with estimates from Cr_2O_3 in grab-samples.

Lambourne and Reardon (1963a) combined samples from individual sheep on a dry-weight basis, and this procedure is preferable when combining for a group of animals. Alternatives are to combine equal weights, volumes or numbers of pellets of fresh faeces, provided there are no obvious secular changes in moisture content; if combining on these bases for a group, the animal(s) with the driest faeces would be over-represented in the composite sample.

Another method of sampling is to collect from defaecations on the sward (Raymond & Minson, 1955). Because the collections will tend to be at random, the risk of bias is minimized; the advantage in using Cr_2O_3 paper rather than gelatin capsules is small, and dosing could perhaps be done once daily. On the other hand, the labour requirement for sampling is high, particularly when large areas are grazed, and though cattle faeces can usually be found fairly easily, this may not be so for faecal pellets from sheep. Care must be taken to avoid contamination of the samples with herbage or other extraneous matter, and there may be changes in Cr_2O_3 : OM ratios owing to weathering of the faeces or insect activity. Faeces from individual animals can be identified if polystyrene particles of distinctive colours are administered at the same time as the Cr_2O_3 (Minson *et al.*, 1960). The particles are about 3 mm diameter, and are used by manufacturers of plastic goods. When ten strip-grazed dairy cows were each dosed twice daily with 20 g particles, the daily identification of 10 to 12 defaecations per cow from the particular colour of particle present, the collection of the samples and marking the defaecations with sawdust after sampling, occupied 1·5 man-hours (Langlands *et al.*, 1963b). An equivalent procedure with sheep drenched twice daily with 15 g coarsely ground particles in 100 ml water on 0·3 ha pastures (Langlands *et al.*, 1963c) took considerably longer. From the results of the experiments with cows it was calculated that for an individual animal given Cr_2O_3 twice daily in capsules and with 70 percent of defaecations sampled, the standard error of an estimate of faecal output over 5 days relative to the output calculated from true mean Cr_2O_3 concentration would be about ± 1 percent, and relative to actual output about ± 6 percent. These errors, which would be approximately halved if sampling was extended over 15 days or slightly increased if sampling efficiency was 50 percent, are about half those associated with grab-sampling.

When pastures are heavily contaminated with the faeces of dosed animals, especially when herbage availability is very low, bias can occur owing to ingestion of faecal material and consequent recycling of Cr_2O_3. When sheep that had never previously been dosed were grazed on such pastures, Young and Corbett (1971a) found faecal Cr_2O_3 concentrations as high as 3 mg/g OM (mean 2·2 mg) compared with the concentrations for dosed sheep of from 10 to 22 mg/g OM, which included the recycling component.

A frequent bias that may arise is the loss of about 3 percent of the Cr_2O_3 in faeces when dried samples are ground in a mill fitted with a cloth collection bag (Stevenson, 1962); the extent of the loss can be determined by comparing results from analyses of faeces prepared in this way with those obtained for subsamples ground in a totally enclosed mill designed to prevent loss of dust. Another procedure is to include in each set of analyses of unknowns, standard faeces samples from penned animals that have been given the same doses of Cr_2O_3 as animals in the field and which have been sampled by the same routine. The Cr_2O_3 doses should also be analysed; the Scherer capsules containing nominally 1 and 10 g are usually found to contain 95 to 98 percent of these amounts.

There are several methods of analysis. A frequently used one is the modification (developed at the Grassland Research Institute, Hurley, 1961) of the method of Christian and Coup (1954). Use of an auto analyzer (Stevenson & Clare, 1963) or an atomic absorption spectrophotometer (Williams, David & Iismaa, 1962) may confer some advantage compared with titrimetry, but chemical preparation before final measurement by these means is the most time-consuming part of the analysis. If $^{51}Cr_2O_3$ is used (Kane, Jacobson & Damewood, 1959) determinations with a γ counter are rapid, but though the half-life of ^{51}Cr is only 27·8 d, the widespread use of this radioactive material would, apart from cost, be inadvisable.

2. Digestibility of grazed herbage

Grazing animals will consume leaf in preference to stem, and green herbage in preference to dead. In consequence, digestibility and nutrient content in grazed herbage are usually greater than in herbage mechanically harvested from a pasture; variation in the magnitude of the difference is also compounded by selection among the range of plant species in the pasture. Methods other than straightforward chemical and digestibility studies on cut, whole plant material must therefore be used to examine the diet of animals at pasture.

(*a*) *Hand-plucked herbage*

If careful observations are made of grazing preferences with respect to the species and parts of plants eaten, material roughly similar in botanical and morphological composition may then be plucked by hand. A modification used in rangeland studies (Cook, 1964) is to pluck material it is known animals will eat, but before they have access to the area, and to collect further material after grazing; comparison of the two sets of samples indicates what has been removed by the animals.

Plucking is so laborious that the collection of quantities sufficient for a conventional digestibility trial with animals is precluded, but sufficient can be obtained for the determination of digestibility *in vitro*, discussed below. A major objection to examination of plucked samples is that the extent to which these differ from the herbage actually grazed is unknown and will contribute an undeterminable bias to the results. On rangelands, even more than with sown grass/legume pastures, the variability in the feed available and in its utilization by the animal is so great that any practicable sampling procedure either before or before-and-after grazing can only provide approximations of the composition of the intake (Langlands, 1974).

(*b*) *Faecal index techniques*

If a representative sample of the feed eaten could be obtained, the concentration of a constituent supposed to be wholly indigestible could be determined and compared with the concentration of the same ‘ internal ’ reference substance in the faeces of animals grazing the pasture. If there is one part reference substance to 100 parts OM in the feed (X percent), and one part to 50 in the faeces (Y percent = 2), then the digestibility of the OM is 50 percent. In general terms :

$$D = 100 \frac{(Y-X)}{Y} \tag{18}$$

Many possible reference substances have been studied, including lignin, the silica that occurs naturally in plants, and plant pigments (chromogen), but none has really satisfied the requirements of resistance to digestion or chemical change in the alimentary tract and ease of determination (Kotb & Luckey, 1972). Some plant silica is absorbed during the process of digestion and other problems arise from soil contamination and from release of silica from stones resident in the rumen. The structure of lignin is not constant and may be altered during passage through the gut, thus limiting the specificity of analytical methods. Reid *et al.* (1950) gave the name chromogen to the pigments that are extracted from herbage by 85 : 15 (v/v) acetone : water. Of these, chlorophyll is the major constituent, and carotene, xanthophyll and porphyrins are also present (Davidson, 1954a, b, c; Deijs & Bosman, 1955). Much of the chlorophyll is excreted in faeces as phaeophytin and there are various other degradation products of the pigments. Deijs and Bosman (1955) recommended that both feed and faeces should be extracted with 85 percent acetone containing 0·3 molar oxalic acid which converts the chlorophyll to phaeophytin and, because the pigments are light-labile (Lancaster & Bartrum, 1954), light should be excluded. Measurements of concentration are made with a spectrophotometer at a wavelength in the approximate range 406-425 nm, the optimum wavelength varying with plant species and stage of growth. There are several reports (e.g. Cook & Harris, 1951 ; Greenhalgh & Corbett, 1960) of faecal recoveries of chromogen departing substantially from the 100 percent level.

Because of the difficulties in obtaining representative samples of grazed feed, techniques have been evolved to estimate composition and digestibility by analysis of material derived directly from that feed, namely faeces. Subsequently, oesophageal fistulation was introduced. However, faecal index techniques are still of value because they do not involve surgical intervention and an estimate of feed digestibility can be obtained for each animal studied.

The concentration of chromogen in the faeces is positively related to digestibility because, in general, the more digestible the herbage, the greater its pigment content and the extent of its concentration in the gut as other substances are preferentially absorbed. Chromogen is less easily determined and is a less precise indicator of digestibility than faecal N (Kennedy, Carter & Lancaster, 1959; Greenhalgh & Corbett, 1960).

The first and still most widely used faecal index substance is nitrogen. Its use stemmed originally from reports of Gallup and Briggs (1948), Raymond (1948) and Lancaster (1949) which drew attention to the fact that much of the nitrogen in the faeces of ruminants has an endogenous origin. This metabolic faecal nitrogen (MFN) includes debris from the alimentary mucosa and of the microbial population, principally from the rumen and of unresorbed secretions into the gut. The quantity excreted is related to the amount of feed DM eaten (Blaxter & Mitchell, 1948) and its concentration in the faeces therefore increases with increasing feed digestibility. Some 5 to 10 percent of the N in faeces is derived from the N originally present in the feed and thus commonly has a true digestibility of 90 to 95 percent. This contribution also enhances the positive relationship between DM or OM digestibility and faecal N percentage, because high herbage digestibility is usually associated with high faecal N contents and the greater absolute amounts of undigested N are excreted in decreasing quantities of faeces.

Nitrogen has the merit that it is easily determined, though there may be considerable losses from faeces during drying (Falvey & Woolley, 1974). However, a given N concentration value does not always indicate the same digestibility value. Either or both of the amounts of MFN and undigested N in faeces may vary without a concomitant change in feed digestibility. What may be described as animal factors include species and physiological state, parasites in the gut which may decrease feed digestibility but increase N loss, and the decrease in digestibility that occurs as the level of intake rises and which is only partially compensated by a change in N excretion. Herbage factors include species and part of plant eaten, season of growth, variation in N content at a given digestibility and N fertilizer application; these are generally more important sources of variability than animal factors in relationships between digestibility (or intake factor) and faecal N percentage, though all may cause serious bias in predicted values of D. For example, Streeter (1969) listed prediction equations from many publications and the digestibility values corresponding to a 3 percent N level in the faecal OM varied from 63 to 78 percent. Faecal N percent has usually been expressed on an OM basis to minimize effects of variation in contamination of the diet by soil.

Milford (1957) found that the digestibility of four subtropical grass species was not satisfactorily predicted from faecal N percentage, but this reflected the very wide range in feed types and composition rather than any inherent difference from temperate species. Relationships for more restricted ranges of tropical and subtropical grasses have been reported by Minson and Milford (1967), Olubajo and Oyenuga (1970) and Jeffrey (1971).

Initial development of the faecal N index technique was based upon the premise that N excretion as g/100 g OM intake is a constant (Gallup & Briggs, 1948; Lancaster, 1949), and Milford (1967) found that intakes of the four subtropical grasses he tested were significantly related to g of faecal N/day. There have been some other reports of the same nature, including one for silage diets (McDonald & Purves, 1957), but such relationships are of limited use because they will hold only within restricted ranges of feeds. For example, they did not hold for a range of Kentucky bluegrass herbages varying in crude protein content from 12 to 19 percent (Forbes, 1949) or for kikuyu grass herbage grown with and without N fertilizer (Jeffrey, 1971). The quantity of faecal N excreted and a number of other variables such as the quantity of faecal OM, dietary N percentage (as determined by analysis of extrusa from animals with oesophageal fistulas), and month of year have been included with N percentage in equations for predicting digestibility (Arnold & Dudzinski, 1963, 1967; McManus, Dudzinski & Arnold, 1967; Langlands, 1969b). The several terms were significant and omission of any would be likely to bias the predictions. Such equations involve considerably more experimental work for derivation than the more usual rectilinear or polynomial regressions of digestibility or intake factor on N percent, but tend to increase the precision of general relationships derived from observations on many types of herbage.

General relationships are imprecise; for example, those reported by Raymond *et al.* (1954) and Minson and Kemp (1951) had residual standard deviations (RSD) from regression of $\pm 5 \cdot 7$ and $4 \cdot 0$ digestibility units, respectively, and in any event, may yield biased estimates of digestibility for particular herbages. It is desirable that the relationship used should have been based on herbage from a pasture similar to, or the same as, the pasture being grazed by the experimental animals. Such ' local ' regressions (Raymond, Minson & Harris, 1956) were obtained by Greenhalgh, Corbett and McDonald (1960) by a continuous digestibility trial procedure. Each time a pasture was grazed, a part of the area was fenced off and harvested progressively to provide standard amounts of fresh feed daily for the animals in the trial. Faecal N percentages and OM digestibilities were determined over consecutive three-day periods; the relationships obtained, all from the same pasture in one year, differed greatly as between spring and autumn growths and to a lesser extent between N fertilizer treatments, but each one had a small RSD of about ± 1 digestibility unit only. The estimates of digestibility were much more precise than if a general prediction equation had been used, but labour requirements for the concurrent continuous digestibility trials are considerable. These will be less if sheep rather than cattle are used, and there is evidence (e.g. Langlands, Corbett & McDonald, 1963) that relationships obtained with two species are similar. There may be significant differences between animals within species, probably linked to variations in MFN excretion, but three animals are adequate in a continuous digestibility trial on one herbage type. The Grassland Research Institute, Hurley (1961) and Minson and Milford (1968) have described equipment for easy management, with minimum labour, of large numbers of sheep in digestibility trials. Further information, including trial procedures, is given in Campbell (1969) and by Harris (1970). Equipment for collecting faeces from cattle is described in references given earlier in this chapter, and by Balch, Johnson and Machin (1962) and Hughes (1963). An alternative and easier procedure is to use Cr_2O_3 as an external indicator. Equation (18) applies; the feed intake of the housed cattle is known, as is the dose of Cr_2O_3 and therefore its concentration in the diet; satisfactorily representative samples of faeces can be obtained fairly readily to determine the contained Cr_2O_3. Precision of faecal N percentage : digestibility equations obtained in this way and by total collection of faeces was similar (Langlands, Corbett & McDonald, 1963), although Cr_2O_3 recovery averaged 97 percent probably owing to the loss of 3 percent of the dose during milling of the faeces.

Obviously a ' local ' regression can only be derived if the herbage on the pasture to be studied is tall enough to cut. If one is derived, errors associated with the regression equation are minimized, but substantial errors may be involved in applying the equations to animals in the field (Grassland Research Institute, Hurley, 1961). Sampling of their faeces is not a major problem because N percentage shows little within-day variation; grab-samples are analysed for N as well as Cr_2O_3. If herbage availability is high and/or heterogeneous in composition, grazing animals may select disproportionately large amounts of one species, or eat more of ' top ' than ' bottom ' growth so that the relation of N percentage to digestibility differs from that obtained with herbage cut and fed (Lambourne & Reardon, 1962; Langlands, 1967b). This may be a smaller problem with cattle than with sheep (Greenhalgh, Reid & McDonald, 1966) because the former are possibly less selective grazers. There is evidence that when the two species graze together on one pasture, there are significant differences between their diet composition (Van Dyne & Heady, 1965; Dudzinski & Arnold, 1973) and in the N/digestibility relationship (Langlands, 1975) though this is not apparent with cut herbage (Langlands, Corbett & McDonald, 1963).

The application error caused by differences in level of intake, and therefore digestibility, between grazing and digestibility-trial animals can be minimized (Langlands & Bennett, 1973a). The approximate levels of intake at pasture in relation to intake levels indoors can be estimated from faecal outputs; digestibilities predicted from the N content may then be adjusted by equation (2) in which the value for D_m will be that obtained indoors whether at the maintenance or some other level of feeding. The possibility of errors due to parasitism should be minimized by regular dosing with anthelmintics.

Large errors may occur if sheep grazing very sparse pasture ingest faecal material. Most of the OM and N in this material will appear in their faeces, thus increasing their faecal OM output but having little effect on the N percentage. Digestibilities predicted for intakes that were actually herbage plus faeces would consequently be similar to those predicted for intakes of herbage alone, though the real digestibility values for the former would be substantially lower. It is probable that this is responsible for the very high estimates of intakes by sheep on pastures with low herbage availabilities reported by Lambourne and Reardon (1963b). Ingestion of soil, resulting in an increase in excretion of N but not in OM, was probably a contributory cause (Young & Corbett, 1972b).

Fresh herbage can be frozen and kept in cold storage for extended periods (Grassland Research Institute, Hurley, 1961) if, for example, it is desired to make detailed studies of herbage at particular growth stages. Digestibility is similar to that of the original fresh material (Raymond, Harris & Harker, 1953) and the technique avoids the considerable changes in chemical and other properties that can result from artificial drying, though a reduction in soluble nitrogen content has been reported (MacRae, Campbell & Eadie, 1975) which will be of importance in some types of studies.

(*c*) *Oesophageal fistula technique*

Since the first use by Torell (1954) of oesophageal fistulation for nutritional rather than physiological studies, the technique has come into wide use. Early problems of establishing and maintaining fistulae have been minimized by improvements in surgical procedures, post-operative management and methods of closure. Surgery on sheep and cattle is described by McManus (1962), McManus, Arnold and Hamilton (1962), Chapman and Hamilton (1962), Chapman (1964), Van Dyne and Torell (1964), Little and Takken (1970) and Hecker (1974), and similar procedures have been used on other species, including reindeer, at the Institute of Arctic Biology at Fairbanks in Alaska. The operation is done in less than 30 min. Short-duration barbiturate anaesthesia only is required for sheep; for cattle, local anaesthesia and sedation with Rompun[1] has been found satisfactory (Langlands, 1975). The operation is less hampered by regurgitation if the animals are not fasted; they should be returned to green pasture or other 'soft' feed, and certainly not to coarse, dry rations. It is advisable to allow 2 to 3 months for recovery before using the animals for routine sampling. Leakage of saliva which may be quite rapid if the plug is removed or accidentally lost, may cause some degree of sodium depletion: this can be offset by providing the animals with a salt lick. Frequent routine examination should be made to check that the plug is in position.

Non-removable acrylic cannulas with a screw cap that can be removed for sampling have been used, but probably the most satisfactory closure is a moulded rubber plug[2] in the form of two L pieces (Plate 18). These are held together externally by one or two strong rubber rings such as those used for castration. A removable flange around the stem at the skin surface may be used to minimize blockage of the oesophagus caused by displacement of the plug. The two pieces of the plug are easily removed for sampling, which should be done at a time when the animals can be expected to be grazing fairly intensively. Prior fasting may sometimes be necessary, though this may affect grazing selection (Arnold *et al.*, 1964). If the fistula is small or badly positioned, or long pieces of feed are swallowed, the extrusa may be only a small and unrepresentative fraction of the material consumed. In this instance the lower part of the oesophagus may be blocked off with a piece of foam rubber, but this is rarely necessary. Extrusa

---

[1] Bayer Pharmaceutical Ltd.

[2] Plugs for sheep are made by T. C. Brown & Co. Pty. Ltd., 181 Clarence St., Sydney, N.S.W.; they are manufactured as T-pieces, and 'split' by cutting the cylindrical stem with a sharp, wet knife. Plugs for cattle are made by Durance Industrial Rubber Supplies Pty. Ltd., 58-64 Leveson St., N. Melbourne, Vic., Australia. The moulds are the respective property of the C.S.I.R.O. Pastoral Research Laboratory, Armidale, N.S.W. and the Victoria Department of Agriculture, Pastoral Research Station, Hamilton, Vic. The manufacturers require an authority to use from these institutions.

PLATE 18. Sheep with oesophageal fistula. The rubber enclosure plug, consisting of two solid L-pieces, is in position in the sheep on the left and has been removed from the sheep on the right

should be collected in a water-tight bag, e.g. heavy gauge plastic, secured around the back of the neck (Plate 19). Usually, samples are collected over periods of 0·5 h and not more than 1 h; the longer the period, the greater the risk that rumen digesta will be regurgitated into the bag, in which event the sample must be discarded. In addition, replacement of the plug will become difficult because the fistula tends to close.

Although the sample is feed actually eaten, it may not truly represent intake. On one pasture there may be variation in composition within and between animals and days (Arnold *et al.*, 1964; Langlands, 1967a, 1969a; Obioha *et al.*, 1970). The fistulated animals should be similar in type, age and previous grazing experience to their companions whose intakes are to be estimated.

The best method of estimating OM digestibility is probably :

(i) Weigh the entire collection of extrusa, and then squeeze it by hand through muslin. Weigh the squeezed solid fraction and calculate the weight of the liquid.

(ii) Determine the DM and OM contents of both fractions. Dry the solid at a temperature not greater than 50°C or freeze-dry to avoid non-enzymic browning reactions in which carbohydrate and protein condense and significantly reduce digestibility values obtained *in vitro*. (See Chapter 5.)

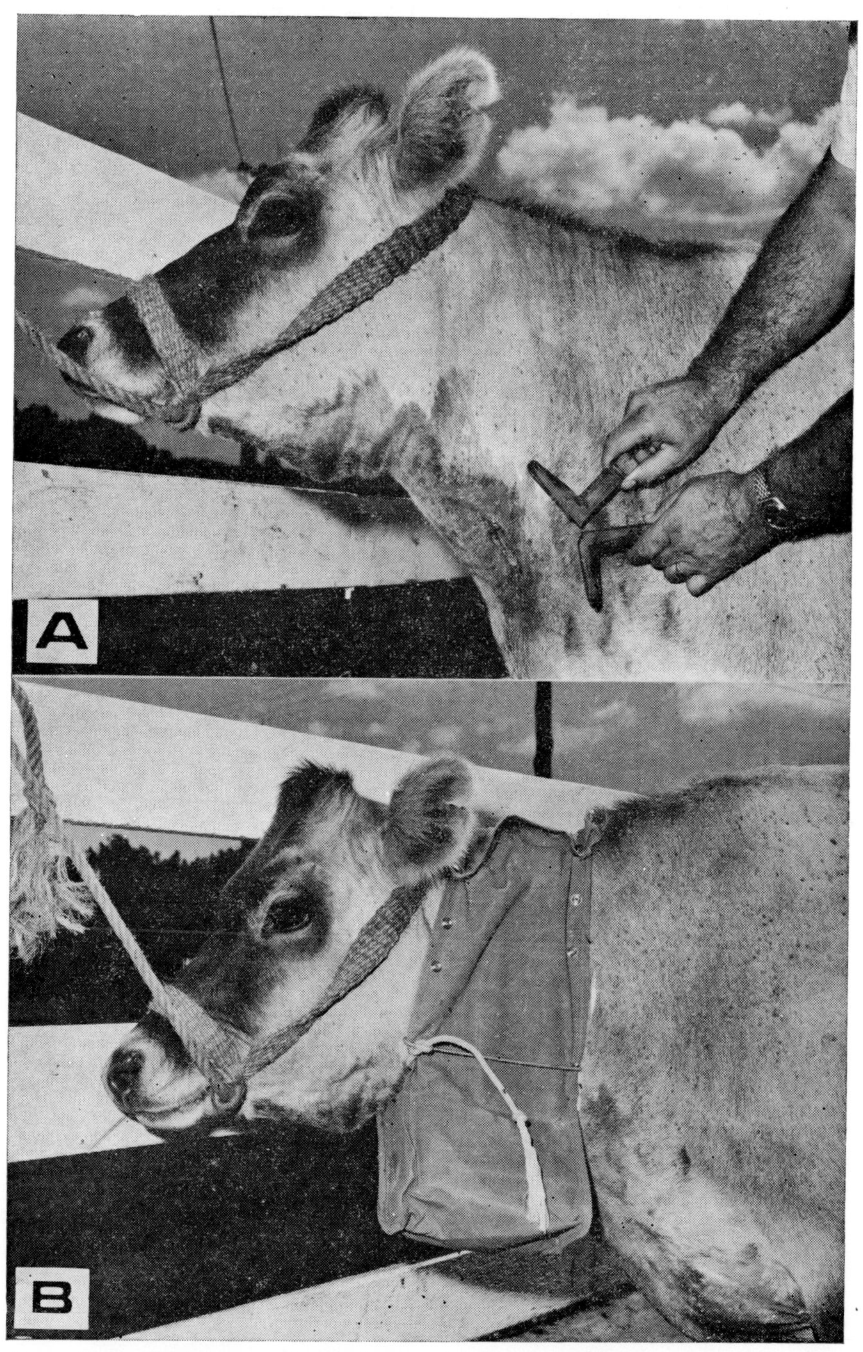

PLATE 19A, B. Herbage sampling by fistulated heifer. A. with plug removed. B. The collection bag has been attached ready for sampling

(iii) Grind the dried solid fraction and determine digestibility *in vitro* by the two-stage procedure of Tilley and Terry (1963), which involves incubation with ruminal fluid and then with pepsin. Various aspects of the technique are discussed by Johnson (1969). Samples of known digestibility *in vivo* must be included in each run so that results for the unknowns can be adjusted to predicted *in vivo* values. The procedure has been adapted for routine use on a large scale (Alexander & McGowan 1966; Minson & McLeod, 1972) and is suitable for browse species as well as pasture plants (Newman & McLeod, 1973). When the digestibility of the samples is 50 percent or less owing to the presence of dry grass residues and other fibrous material, Wilson, Weir and Torell (1971) recommended use of the modification of the two-stage procedure suggested by Van Soest, Wine and Moore (1966); the second stage uses a neutral detergent and more vigorous washing to improve the separation of indigestible cell wall materials from extraneous substances.
The use of ruminal fluid in the first stage does not represent an attempt to mimic digestion *in vivo*; the fluid is used as a source of a wide range of cellulolytic and other enzymes (Corbett, 1969). There is some inconvenience in maintaining animals with ruminal cannulas as donors, and work by Jones and Hayward (1975) and McQueen and Van Soest (1975) with fungal cellulase indicates this as a possible alternative to ruminal fluid.

The organic matter in the liquid fraction is partly of salivary origin and partly water-soluble carbohydrates and protein and suspended matter from the herbage which has been released during mastication. If it is assumed all the organic matter is of feed origin, digestibility will be overestimated by 1 or 2 percentage units; it will be underestimated by 3 to 4 units if the liquid fraction is assumed to be entirely saliva and is discarded (Langlands, 1975). Attempts have been made (Langlands & Bowles, 1973) to estimate saliva in the liquid from the tritium present after TOH was injected into the animal and the specific radioactivity of the saliva was determined, but these were unsuccessful. Salivary OM has been taken to be 358 mg/100 ml liquid fraction (Grimes, Watkin & May, 1965), but Langlands (1975) found that it was better calculated as equivalent to 6·3 percent of the organic matter in the squeezed solid. Non-salivary OM in the liquid is assumed to be 100 percent digestible. Digestibility of the sample of grazed herbage can then be calculated by the following expressions :

$$D = \frac{(A \times B) + (C - 3{\cdot}58\ E)}{B + (C - 0{\cdot}358\ C)} \times 100 \qquad (19)$$

$$\text{or} \qquad D = \frac{(A \times B) + (C - 0{\cdot}063\,B)}{B + (C - 0{\cdot}063\,B)} \times 100 \qquad (20)$$

A is the predicted digestibility *in vivo* of the OM in the solid fraction,
B is the weight of OM in the solid fraction,
C is the weight of OM in the liquid fraction, and
E is the volume (l) of the liquid fraction.

The best procedure, however, is to determine with penned animals the digestibilities *in vivo* of the types of feed that are of interest, and establish by regression analysis the relationship with *in vitro* values determined on extrusa samples obtained during the same study.

When the collection period is prolonged because grazing is spasmodic, the extrusa may include disproportionately large quantities of saliva. If the amount of OM in the liquid fraction is greater than 25 percent of that in the solid, it may be preferable to discard the sample, but such a distribution of OM is not uncommon in 'good' samples. Langlands (1975) has suggested that the digestibility of the intake (D_i) be estimated from the digestibility of the solid fraction only (D_s) by the following equation :

$$D_i = 38{\cdot}5 + 0{\cdot}00695\,(D_s)^2 \qquad (\text{RSD} \pm 2{\cdot}5) \qquad (21)$$

This equation was derived from observations with penned sheep given a wide range of temperate pasture herbages and should not be used for other forages, particularly those of low digestibility.

The precision of the technique is indicated by the residual standard deviations from regression equations relating digestibility of extrusa to the respective measured digestibilities *in vivo* of the feeds, and by the error term in analyses of variance of results for extrusa obtained on several days from a number of fistulated sheep on a pasture. These SD values are generally ± 3 to 4 digestibility units, or as coefficients of variation about ± 4 to 6 percent, and because the SD between animals or days are usually smaller, Langlands (1975) suggested that error in the technique is attributable mainly to the sample processing and *in vitro* procedures. The estimates may be adjusted to allow for level of feeding effects as already described.

In general, estimates of the digestibility of grazed herbage obtained by the oesophageal fistula technique are rather less precise than those that can be obtained by prediction from faecal N percentage. However, the latter technique carries a greater risk of bias; this bias can be very large and, unlike inaccuracies in results obtained from extrusa, the main causes do not arise from faulty laboratory procedures that can be detected and corrected.

3. Chemical and botanical composition of grazed herbage

The N contents of oesophageal extrusa are little different from and can be used to predict N in feeds (Langlands, 1966b; Marshall, Torell & Bredon, 1967; Little, 1972). Sampling errors are similar in type to those associated with the determination of digestibility, and may be greater in magnitude because diurnal variation with N appears to be larger (Langlands, 1967b). Some other organic constituents of the feed, such as structural carbohydrates, can also be estimated (Torell, Bredon & Marshall, 1967; Hoehne, Clanton & Streeter, 1967), but the mineral content of the saliva that contaminates the extrusa precludes estimation of many dietary minerals. Calcium concentration appears to be least affected (Langlands, 1966b; Hoehne, Clanton & Streeter, 1967); potassium concentration is reduced by leaching, while sodium, phosphorus and chlorine are substantially increased and these elements cannot be estimated, even though attempts are made to estimate salivary contamination with TOH (Langlands & Bowles, 1973). Little (1975) found with cattle that dietary levels of Ca, S, Cu and Mg, but not of Mo, Mn and Zn, could be predicted with an error of less than ± 9 percent. Concentrations of Ti were accurate and may be a useful indicator of the extent to which grazed herbage is contaminated by soil.

The botanical composition of grazing intake can be estimated by examination of extrusa. Heady and Torell (1959) described the use of a binocular dissecting microscope to identify plant fragments ' hit ' by a cross-hair reference point during traverses of thinly-spread extrusa. The technique, which is analagous to the point-quadrat method for studying pastures (see Chapter 3), was further developed by Harker, Torell and Van Dyne (1964), who described and illustrated it in detail. Hamilton and Hall (1975) showed that the results obtained may give substantially biased estimates of composition by weight because weight per unit area, i.e. per hit, varies among plant species. They determined weight per unit area constants for the materials they studied, and recommended this be done routinely to allow adjustments of observations to a weight basis. This procedure was more accurate and less laborious than hand separation of extrusa (Hall & Hamilton, 1975). The investigator has to become familiar with the morphology under magnification of the various plant species and plant parts available for grazing. Such difficulties are least when the intake consists mainly of two dissimilar species, such as a legume and a grass, and diminish as experience is gained.

Methods for estimating botanical composition by examining samples from the rumen of live animals have only limited applicability and value (Van Dyne & Torell, 1964). They can be used to gain general information on the feed preferences of wild animals that have been recently killed for this or some other purpose (Hofmann & Stewart, 1972). Another method is to examine the faeces and identify fragments of cuticularised epidermis from the plant species ingested (Croker, 1959; Hercus, 1960; Storr, 1961). Although it has been claimed that a species contributing more than 5 percent by weight of the diet can be indentified and quantified (Stewart, 1967; Free, Hansen & Sims, 1970), the epidermis of some plants may not survive digestion to an extent sufficient for identification, thus causing bias in the results (Slater & Jones, 1971). The procedures for preparation and examination of faeces and for establishing frequency-to-density correction factors are laborious (Sparks & Malachek, 1968; Williams, 1969) but the technique has a useful role in studies of wildlife ecology (e.g. Stewart & Stewart, 1970; Griffiths, Barker & MacLean, 1974).

The phenolic substance orcinol is present in heather (*Calluna* sp.) and is quantitatively excreted in sheep urine. Martin, Milne and Moberley (1975) have shown, that if determinations are made of the orcinol concentration in oesophageal extrusa from sheep grazing heather moors and of the total feed intake, the quantity of heather eaten and its proportion in the diet can be estimated from urinary orcinol excretion. A similar technique could be used in grazing studies on other mixed plant communities if one or more plants contained a substance eliminated in urine.

4. Grazing intake

It will be seen from equation (3) that estimation of grazing intakes from faeces measurements actually involves determination of the indigestibility (100-D) rather than the digestibility of a feed. In consequence a small error in the latter, say 1 percentage unit, results in errors in the estimated intake amounting to 3·3, 2·5 and 2·0 percent at digestibilities of 70, 60 and 50 percent, respectively. As has been seen, substantially greater biasing errors than 1 unit can occur in estimates of digestibility, and thus intakes. When animals graze together on the same pasture, information on their relative rather than absolute intakes can be obtained by measuring only the faeces output, provided that the diets selected by the individuals are similar.

The greatest precision that might be obtainable for estimates of digestibility, either predicted from local faecal N percentage equations or obtained by intensive application of the oesophageal fistula technique, could be of the order of ± 1 percent, and for indigestibility about ± 3 percent (when $D = 75$ percent). The error of faeces output estimates obtained by an efficient sward-sampling routine might be as low as ± 5 percent, and probably no less for measurements made by collection because of end-of-period errors (Blaxter, Graham & Wainman, 1956) when faeces that should (or should not) have been collected are voided after (or before) the collection bags are removed. These errors in the two measurements, if independent, combine to give a total error of $(3^2 + 5^2)^{0.5} = \pm 6$ percent for estimates of intake. More realistic values in many situations for the component errors will be ± 10 to 12 percent, giving a combined error of about ± 16 percent.

In addition there will be real differences in intake between animals of similar type, etc. grazing together on a pasture, with a coefficient of variation of the order of ± 7 to 14 percent (Corbett, 1969). The implications for the design of experiments involving estimation of intakes when the coefficient of variation for the estimate may be as much as ± 20 percent, have been discussed by Ulyatt (1972) and will also be considered at the end of this chapter.

There are some other techniques for estimating grazing intakes. Measurement of the quantity of herbage present on a pasture at a particular time are described in Chapter 4. If such estimates are made by mechanical, electronic or visual means or by a combination of these at the start of a period of grazing and again at the end, the loss of herbage is taken to represent the total intake of as many animals as are in the group. The total intake can be described in terms of the quantity of DM or OM and, if ' before and after ' herbage samples are taken for analysis, of crude protein, etc. also; the problems in these determinations are those discussed earlier (hand-plucked herbage). In addition, intakes will either be underestimated if the growth of herbage during the grazing period is ignored, or will be overestimated if trampled herbage, which is not easily seen or measured, is not included in the ' after ' yield estimate, or if there is grazing by feral animals, and to the extent that herbage is lost by decomposition, insect activity and wind. Probably the most favourable conditions for application of the technique are intensive grazing of fairly uniform pasture, with the animals moved daily to completely separate areas, for example strip grazing with a rearward fence. Problems and errors multiply as the size of the area and the duration of the grazing period increase; at best, the errors are likely never to be less than substantial (Corbett & Greenhalgh, 1960; Campbell, 1969) and the method is not often used.

Another technique proposed by Erizian (1932) depends on observations of short-term changes in liveweight :

$$\text{Intake} = (\text{Wt}_2 + \text{F} + \text{U} + \text{I}) - \text{Wt}_1 - \text{L} \qquad (22)$$

where Wt_2 and Wt_1 are liveweights respectively after and before a period of grazing,

F and U are the weights of faeces and urine voided during that period,

I is the ' insensible loss of weight ', principally evaporative loss and

L is the weight of water drunk.

This series of measurements provides only an estimate of the intake of fresh herbage of unknown DM content etc., and so tells little of pasture productivity or animal nutrition. Allden (1969) was able to gain more information in special circumstances. He studied sheep in a Mediterranean environment under conditions in which the pasture had matured and dried off and in which there was virtually no change in the composition of the senescent annual and perennial plants over a long period. Digestibility was determined with housed sheep given cut herbage. The sheep outdoors were allowed to graze for 2·5 h twice daily without access to water, and faeces and urine were collected. Liveweights were measured to the nearest 10 g by slinging the animals from a spring balance, and adjustments for insensible loss were made from observations on the liveweight changes of similar fully harnessed sheep not allowed access to grazing during the 2·5-h periods. Faeces outputs over 7 to 10 days were measured by collection and combined with digestibility values from the housed sheep to give digestible intakes.

5. Intake of supplementary feeds and water

The grazing intake of animals is reduced by the consumption of supplementary feeds, though their total feed intakes may be increased. The supplements are thus in reality partial substitutes. Holmes and Jones (1964) reviewed a number of experiments with dairy cows on temperate pastures and found that the increase in total intake (I, kg) per kg of concentrate eaten varied with the digestibility of the OM in the herbage grazed :

$$I = 2{\cdot}8 - 0{\cdot}034D \qquad (23)$$

Holder (1962) expressed the extent of substitution as a coefficient of efficiency of supplementation (Q) :

$$Q = \frac{(P + S) - P}{S} \qquad (24)$$

where (P + S) is the total digestible OM or energy intake of supplemented animals, P is the pasture intake of those not receiving supplement, and S is the quantity of supplement eaten. If Q = 1·0 there has been no depression in grazing intake, and if Q = 0·1 then it has been reduced by an amount equivalent to 90 percent of the digestible OM or energy in the supplement. Equation (23) indicates that these values of Q would occur at D values of 53 and 82, respectively, but the extent of the reduction varies with herbage availability and, in general, is greatest when there is abundant pasturage (Langlands, 1969c).

When a known amount of a supplement is consumed by grazing animals, their herbage intake can be estimated if it is assumed that the digestibilities of the herbage and of the supplement, previously determined, are the same whether consumed alone or together. Alternatively, the two feeds may be given together in various proportions to animals in a digestibility trial, and the separate digestibilities estimated from the partial regression coefficients in equations relating intakes of the two dietary components to faecal outputs (Langlands, 1969c). At pasture, the digestibility of the herbage grazed is estimated by the oesophageal fistula technique and faecal output values are reduced by the quantities of indigestible residues calculated to be derived from the supplement. An assumption that there are no associative effects on digestibility may be invalid when the supplement provides a nutrient, for example N, that is deficient in the herbage.

When a supplement, or water, is offered *ad libitum* the quantities consumed by individual animals can be estimated if it is labelled with an indicator such as TOH. This technique was devised by Leng *et al.* (1975); it has been used to study the intakes of urea-molasses mixtures by sheep and cattle (Nolan *et al.*, 1974, 1975) and by Langlands and Holmes (1975) to estimate intakes by cattle of mixtures containing bloat prophylactics.

6. Other measurements

(*a*) *Ruminal and post-ruminal digestion*

Ruminant animals gain a substantial part of the DE in their feed in the form of steam-volatile fatty acids (VFA), principally acetic, propionic and butyric. Farrell, Leng and Corbett (1972a) estimated the rates of production of VFA in the rumen of adult grazing sheep from the concentration in samples of ruminal fluid withdrawn automatically and continuously *per fistulam* (Farrell, Corbett & Leng, 1970). The prediction equation used (Corbett, Leng & Young, 1969) was derived from measurements of VFA production rates by a radionuclide dilution technique, and of VFA concentrations made on similar animals indoors and at pasture. It was assumed that the heats of combustion of the VFA produced represented 55 percent of DE, and that ME intakes were 82 percent of DE. The results showed good agreement with ME intakes estimated by other means, as did those obtained by Langlands (1975) in a digestibility trial with housed sheep given known amounts of feed.

In a development of this technique, Corbett *et al.* (1975a) determined the extent of post-ruminal digestion in grazing lambs with the aid of two markers, ^{103}Ru labelled tris (1,10- phenanthroline) ruthenium (II) chloride (Tan, Weston & Hogan, 1971) and the ^{51}Cr complex of ethylenediaminetetra-acetic acid (Downes & McDonald, 1964); the ^{103}Ru-P is strongly absorbed by particulate matter and the ^{51}Cr EDTA remains in solution, thus marking the liquid phase

PLATE 20. Lamb with harness (Royal, 1968), faeces collection bag and rumen cannula. A pump (Corbett *et al.*, 1976b) over the shoulder infuses liquid from the pack on the rear shoulder through a cannula into the rumen and withdraws ruminal fluid continuously through a second line into a pack on the other shoulder.

of the digesta (MacRae, 1974). The markers in solution were infused into the rumen by pumps (Corbett *et al.*, 1976b) carried by the lambs (Plate 20). Analysis and radioactivity assays were made on abomasal digesta, taken through a permanent cannula, and on the faeces. The amounts of OM and its constituents leaving the abomasum and excreted were calculated (Hogan & Weston, 1967) and, by difference, the absorption from the hind gut. The pumps also withdrew ruminal fluid continuously, and VFA productions were estimated with an equation relating these to pool sizes in the rumen (VFA concentration per l × ruminal fluid volume in l). Digestible OM intakes (digested in rumen plus digestion in hind gut) were obtained without making the assumption that VFA represented a constant fraction of DE, and information was gained on the quantities of various nutrients absorbed by the lambs. It is possible to determine Ru and Cr by X-ray fluorescence spectrometry (MacRae & Evans, 1974), or perhaps both by atomic absorption spectroscopy, thus obviating the need to use radioactive isotopes.

The technique cannot be widely adopted and, generally, should only be used to study special problems of nutrition. It does yield estimates of grazing intakes that are independent of, and so are a valuable check on, estimates obtained by other means such as the oesophageal fistula/faecal output techniques; in the study with lambs, values by the two different methods were in good agreement.

Methods for cannulation of various parts of the digestive tract have been described by Hecker (1974).

(*b*) *Methane and urine*

The energy of methane and urine is about 18 percent of the DE, but more exact information may be required.

If VFA production is estimated, methane production can be calculated stoichiometrically (e.g. Leng, 1970). It can be determined in the field with tracheotomized animals, as discussed below. Rates of gas production in and elimination from the rumen are of importance in bloat; these have been measured by a radionuclide dilution technique using tritiated methane (Bryant, Murray & Leng, 1973) but so far only with housed animals.

Knowledge of urine outputs may be needed to examine recycling of plant nutrients, especially N and K, or to examine losses of energy and nutrients by the animal. Some desert range plants in the USA contain substantial amounts of essential oils which are excreted in the urine and as a result ME is substantially less than 82 percent of the DE (Cook, Stoddart & Harris, 1952). Equipment for collecting urine from steers has been described by Border, Harris and Butcher (1963) and from grazing wethers by Sears and Goodall (1942), Cook, Stoddart and Harris (1952), Bassett (1952), Beeston and Hogan (1960), and Engels and Hugo (1964). The last mentioned employed a closed plastic container with a simple non-return valve which prevents spillage and which, with modification, could probably be used equally satisfactorily with steers; the pump described by Corbett *et al.* (1976b), could be used to withdraw a sample of the urine to another container on the animal's back, the remainder of the urine being pumped to waste. For cows, catheterization (Cunningham, Frederick & Brisson, 1955) or use of a urine separator (Van Es & Vogt, 1959) could be combined with a sampling device (Lesperance & Bohman, 1961) to determine total output, as has been done with ewes by Chambers *et al.* (1975). The bladder of sheep and cattle has been catheterized directly by entry through the flank of the animal by J. J. Lynch of the CSIRO Pastoral Research Laboratory, Armidale, NSW.

All these methods present problems. It was hoped that creatinine, which is excreted in the urine in approximately constant daily amounts that are related to body mass (Mitchell, 1962; Van Niekerk *et al.*, 1963), could be used as an index substance (Butcher & Harris, 1957). Unfortunately, the ratios of creatinine to N and other substances in urine vary considerably during the day; estimates of daily outputs of N and other substances by reference to the amounts of creatinine in the urine sample and excreted in 24 h are, in consequence, subject to considerable errors (Langlands, 1966a).

If an adequate sample of urine is obtained, energy value is more easily calculated from the N content (Paladines *et al.*, 1964a; Street, Butcher & Harris, 1964) though less accurately than if determined by bomb calorimetry (Blaxter, Clapperton & Martin, 1966).

(*c*) *Blood*

Devices exist that permit continuous sampling or single automatic collections of jugular venous blood (Farrell, Corbett & Leng, 1970; Corbett *et al.*, 1976b). Chemical analyses may assist definition of nutrient deficiencies, especially of minerals. Haematological studies, apart from pathological examinations that may be required, have little value in the context of this book; for example, a low haematocrit value will only confirm what will already be obvious—that the animal if otherwise healthy is undernourished (e.g. Farrell, Leng & Corbett, 1972a).

(*d*) *Energy expenditure*

The ' classical ' method of determining the energy expenditure of animals, i.e. the measurement of their oxygen consumption and carbon dioxide production, has been used on grazing sheep (Corbett, Leng & Young, 1969; Young & Corbett, 1972a). These were fitted with a re-entrant tracheal cannula (Young & Webster, 1963) that separated exhaled gas (including methane) from inspired air. The technique avoided the use of a mask which would, of course, prevent grazing. A similar technique was proposed for cattle (Flatt *et al.*, 1958). The method could be improved by the use of small portable transducers now becoming available for measuring gas composition and gas flow, and perhaps by newer surgical techniques (Borrie, Redshaw & Dobbinson, 1973). It is probably of most use when short-term changes in energy expenditure are to be studied, for example, responses of animals to rain squalls and expenditures during grazing compared with resting.

Several other methods are discussed in a publication edited by Gessaman (1973) including heart or respiratory rates, and turnover rates of various labelled substances in the body, but these are generally unreliable with the exception of doubly labelled water, $D_2{}^{18}O$ (Lifson, Gordon & McClintock, 1955). Indeed, this last method is perhaps the most satisfactory from a technical standpoint because it allows the energy expenditure of animals to be determined for long periods while they are undisturbed in their natural environment. The turnover rates of the hydrogen and oxygen in body water are estimated by reference to the introduced deuterium and oxygen-18 respectively. The rate for oxygen is greater because it is lost from the body mainly in CO_2 as well as water, while hydrogen is lost mainly in water; comparison of the two rates allows estimation of CO_2 production and thence, with less precision, O_2 consumption. Unfortunately, the present cost of the labelled water virtually prohibits its use on other than small animals such as rats and pigeons.

A method that has been used successfully on grazing cattle and sheep and some other animals is the carbon dioxide entry rate technique (Young, 1970; Corbett *et al.*, 1971; Young & Corbett, 1972a; Whitelaw, 1974). Labelled carbon dioxide, as $NaH^{14}CO_3$, is infused at a constant rate (nCi/h) into the subject; after some hours to allow for equilibration, body CO_2 is obtained by acidifying samples of blood, urine or saliva (Engels, Inskip & Corbett, 1976) taken at intervals or continuously during 24-h periods, and its specific radioactivity (nCi/*l*) is determined. CO_2 entry rate $\left\{\frac{(\text{nCi/h}}{\text{nCi}/l} = l/\text{h}\right\}$ is calculated and from this is estimated the energy expenditure (kJ/h) by an equation previously derived with similar animals. The equation relates CO_2 entry rates to energy expenditure determined simultaneously from respiratory gaseous exchanges. The residual standard deviations, as coefficients of variation, for a number of such relationships have been in the range ± 8 to 16 percent.

VII. ANIMAL BEHAVIOUR

Observations on grazing behaviour and on social, sexual and maternal behaviour may illuminate interactions between climate, soil, plant, animals and management, and complement measurements of the effects of changes in any of these factors on the productivity and stability of grazing systems. Information so gained can indicate ways to improve animal production, reproduction and survival. For example, studies with housed animals show that feed intake is affected by many plant and animal factors, but at pasture their relative importance is modified by the sward structure which can affect rate and quantity of intake as shown by observations on bite size and rate of biting of cattle grazing subtropical pastures (Stobbs, 1973a, b). Other examples are: effects on reproduction rate of variation in libido between males, and in activity between females during oestrus; effects of climate, and local modification by topography and natural or artificial shelter, on plant growth, grazing behaviour, and survival of the newborn; transference of plant nutrients in faeces and urine to a ' camp ' habitually used by gregarious animals, thus causing depletion of the greater part of the grazing area; changes in the vegetation on rangelands in relation to the availability of drinking water and thence ranging activity by stock, or in relation to the feed preferences exhibited by grazing and browsing of various species on a given area. These and other matters, and the implications and consequences for the management and productivity of plants and animals, are discussed in numerous publications; those by Hafez (1969), Lynch and Alexander (1973) and Kilgour (1974) are particularly useful.

In general, observations can assist in understanding why changes in the naturally dynamic, grazed communities of plants take particular directions; objective descriptions of such changes require application of techniques discussed in this and other chapters. Their application may be assisted by knowledge of behaviour. For example, collection from animals with oesophageal fistulas is easiest during a normal grazing period; and the objectives of studies with grazing animals might not be satisfactorily fulfilled if, in their design, such matters as size of animal group or uneven redistributions of plant nutrients in excreta were ignored (see Animal Management, this chapter).

Patterns of animal dispersion over grazing areas are far from uniform, and can reflect variation in the distribution of feed and its utilization, social interactions among the animals and interactions between the animals and the total environment. Quantitative descriptions of behaviour involve measurements of the frequency of occurrence of activities or displays, time intervals between these occurrences and in relation to the presentation of stimuli, the intensity of activity and how this changes with time owing to such factors as fatigue, satiation and habituation, and the rate of recovery in response behaviour after such factors have supervened.

Features of grazing behaviour that are of interest include times spent grazing, the total number of bites and bites per minute, rumination times and the number of boluses and bites per bolus, the number and frequency of drinks and volumes drunk, times spent lying down, times spent idling or ' loafing ', and distances walked. Such studies are exemplified by the detailed reports of Hancock (1950; 1954a, b) and Hancock and McMeekan (1954); these were the outcome of observations made at intervals of one to five minutes over numerous 24-h periods on the activities of each of 20 individually identified cows. The equipment required for such work includes shelter at a suitable vantage point for the observer(s), a watch with a sweep second hand, binoculars, and a form on which the various activities can be recorded simply and on which, occasionally, comments can be written. At night it is often necessary to walk around the animals and observe with the aid of an electric torch, used sparingly to avoid disturbance. Castle, Foot and Halley (1950) made some observations with the aid of infra-red equipment instead of visible light.

Dispersal of animals in small areas can be plotted on paper marked with a grid with reference points corresponding to prominent natural features (trees, gullies, ridges) or markers placed on fences, etc. A similar technique can be used on larger areas where the topography allows (Kilgour, Pearson & de Langen, 1975), but where the entire area cannot be viewed from one point, observations may be made from a vehicle driven along transects (Lynch, 1974) or by aerial photography (Dudzinski & Arnold, 1967). Recording animal movements on paper marked with a grid can provide information on distances walked. An alternative is to harness the animals to a ‘ range meter ’ (Cresswell & Harris, 1959), which is a towed wheel fitted with a trip-counter to record revolutions.

Reduction of the data obtained from behaviour studies to provide clear succinct descriptions of the important behavioural traits can present problems. The results can be presented in diagrams and charts as well as tables (e.g. Hancock, 1954a, b), and though dispersal patterns can be shown in such ways if indices of distribution are defined (Dudzinski, Fahl & Arnold, 1969), it is probable that increasing use will have to be made of specialized computer programmes (Kilgour, Pearson & de Langen, 1975). Greater use of techniques of pattern analysis (Williams, 1975 and Chapter 8) is likely.

Various other measurements may be required for meaningful interpretation of the observations of behaviour. These measurements may include ordinary meteorological records, duration and intensity of rainfalls, special records of air movement at animal height at various sites varying in topography, and of temperatures and solar radiation at these sites (Meteorological Office Great Britain, 1956). Measurements of grazing intakes associated with observations on the number and rate of grazing bites allow calculation of the variation in intake per bite between types of pasture (Stobbs, 1973a, b), and variation in herbage availability is reflected in time spent grazing which in turn affects energy expenditure by the animal (Farrell, Leng & Corbett, 1972b; Young & Corbett, 1972a).

The fascination in watching animals tends to pall when this is continued over long periods during both night and day and, as it should be, in foul as well as fine weather. There is a considerable range of devices to lighten the burden of observation. Rutter (1968) used time-lapse photography to study behaviour of livestock at pasture, and Squires, Daws and Bawden (1969) used automatic photographic equipment actuated mechanically by animal movements, and which could be triggered by interception of light falling on a photoelectric cell. Videotape recorders have the advantage that the tape can be re-used when no more playback is required.

Allden (1962) attached clocks[1] to sheep. These clocks incorporate a stylus sensitive to movements made in grazing and ruminating, and which record these on a time chart. This equipment is used quite widely; Stobbs (1970) used it to record automatically the duration and periodicity of grazing by dairy cows. An electronic device for measurement of the grazing time of cattle has been described by Jones and Cowper (1975) (Plate 21).

Transducers sense events such as mechanical movements, changes in pressure, temperature or other physical, physio-chemical or chemical phenomena, and convert these to an electrical signal. The signal, suitably amplified, is transmitted to a recorder system. The transmission may be by wires to a recorder borne by the animal, or by radio waves directly or through a local amplifier-repeater radio transmitter to a receiving and recording station. These processes are referred to as telemetry, a term denoting measurement at a distance by electrical means; this applies whether transducers and recorder are connected by wires or whether (as in the case of radio-telemetry) they are not.

A variety of telemetric techniques has been or could be applied to studies of grazing behaviour. These include an animal-borne recorder of jaw movements sensed by microswitch, and grazing position sensed by mercury switch (Stobbs & Cowper, 1972); radio-transmission of jaw movements sensed from electric potentials in the masseter muscles (Nichols, 1966); a similar system providing information on tooth contact (Kavanagh & Zander, 1965); a small pressure transducer embedded in a tooth to sense biting (Kydd & Mullins, 1963); animal-borne equipment to record a variety of activities (Canaway, Raymond & Tayler, 1955); and a system for locating cattle in very large areas (Petrusevics & Davidson, 1975).

If studies are to be made on wild animals such as bears, giraffe, deer, or birds, aquatic mammals or fish (Mackay, 1970) there is often no satisfactory alternative to radio-telemetry. With domestic livestock, Bligh and Heal (1974) stated that this technique remains a largely unfulfilled hope. The high cost of developing some systems, particularly those for recording several biological functions concurrently in a number of animals, has been barely repaid by the rather small volume of reliable and useful information obtained. Good records are still difficult to obtain and it is common experience that as much or more time is spent on solving technical problems as on the collection of worthwhile results.

These views would be echoed by many who have been associated with radio-telemetry systems. Certainly " the promise of a means to study biological

---

[1] 'Vibracorders' originally designed to record the duration of movements by vehicles, and manufactured by Kienzle Apparate G.m.b.H., Villingen im Schwarzwald, W. Germany.

PLATE 21. Compact light-weight electronic grazing clock fitted to halter using batteries and a mercury switch which closes when the head is lowered. Grazing time is registered on an impulse counter which can be read from a short distance (Jones & Cowper, 1975)

function under . . . natural conditions . . . is too important to be ignored " (Bligh & Heal, 1974), but those proposing to enter this field should be cautious. They should appraise the situation with respect to what measurements are required, the degree of accuracy desired, and whether wired connections could be used instead of a radio link, which is much more complicated and expensive.

VIII. CONCLUSION

The discussion of the various techniques and their application has dwelt upon the labour requirements, which are often high, and the errors, particularly biasing errors, which may be substantial. The intention has not been to discourage use of the techniques, but rather to encourage a realistic approach from which will stem effective studies of important aspects of the grazing ecosystem.

The precision of the techniques has been described in terms of coefficients of variation. This information can be used to maximize the chance that an experiment will yield decisive results, or can help avoid expenditure of resources on one that is unlikely to have a meaningful outcome. Suppose that the coefficient of variation for an estimate of grazing intake is 15 percent, then Table 1A shows that if six observations are made for each of two grazing treatments there is a 90 percent chance ($\beta = 0 \cdot 90$) of finding a difference between the treatments significant at $P < 0 \cdot 05$ if the difference is as large or greater than 31 percent. If the experimenter is inclined to accept a slightly, instead of substantially, better than even chance, $\beta = 0 \cdot 60$ instead of $0 \cdot 90$, then (Table 1B) the minimum difference would be about 21 percent. Tables 2A and 2B are complementary; where the CV and the likely extent of a difference can be assessed, these show the numbers of observations required to establish the difference significant at $P < 0 \cdot 05$ for β of $0 \cdot 90$ and $0 \cdot 60$. Interpolations will provide a guide for other β etc., and perhaps a general indication when more than two treatments are to be compared, or Pearson and Hartley (1972) may be consulted.

When the Tables are used in planning experiments that involve grazing animals, special consideration must be given to what is meant by an ' observation'. Particular care is required when it is known or suspected that there will be interactions between soil, plant and animal. These circumstances will occur quite frequently in grazing experiments and then each area of pasture plus the animals that graze it must be taken as a single experimental unit; treatment effects on the animals must be assessed from mean performances and not by analysis of individual animal results. There is a helpful discussion of this and related topics by the Grassland Research Institute, Hurley (1961), Morley and Spedding (1968) and in Bofinger and Wheeler (1975) (see also Chapter 6), but in any event the design of each experiment should first be discussed with a statistician.

CONDITIONS FOR DETECTING A SIGNIFICANT DIFFERENCE ($P < 0{\cdot}05$) BETWEEN TWO GROUPS EACH OF m OBSERVATIONS WITH PROBABILITY β

TABLE 1. *MINIMUM PERCENT DIFFERENCES for various nos. of observations (m) and values of coefficient of variation (CV%)*

TABLE 2. *MINIMUM NOS. OF OBSERVATIONS for various percent differences (D) and values of coefficient of variation (CV%)*

Tables 1A and 2A $\beta = 0{\cdot}90$

m / CV	4	5	6	7	11	16	31
2	*5·5*	*4·7*	*4·2*	*3·8*	*2·9*	*2·4*	*1·7*
5	*13·9*	*11·7*	*10·4*	*9·4*	*7·3*	*5·9*	*4·2*
10	*27·7*	*23·4*	*20·8*	*18·9*	*14·5*	*11·9*	*8·4*
15	*41·6*	*35·2*	*31·2*	*28·3*	*21·8*	*17·8*	*12·6*
20	*55·4*	*46·9*	*41·6*	*37·8*	*29·1*	*23·7*	*16·7*
25	*69·3*	*58·6*	*52·1*	*47·2*	*36·3*	*29·6*	*20·9*

D / CV	2	5	10	15	20	25
2	*24*	*5*	Less than 4			
5		*24*	*7*			
10			*24*	*11*	*7*	*5*
15	Greater than 31			*24*	*16*	*11*
20					*24*	*16*
25						*24*

Tables 1B and 2B $\beta = 0{\cdot}60$

m / CV	4	5	6	7	11	16	31
2	*3·7*	*3·2*	*2·8*	*2·6*	*2·0*	*1·6*	*1·1*
5	*9·4*	*8·0*	*7·0*	*6·4*	*4·9*	*4·0*	*2·9*
10	*18·7*	*15·9*	*14·0*	*12·9*	*9·9*	*8·1*	*5·7*
15	*28·1*	*23·9*	*21·1*	*19·3*	*14·8*	*12·1*	*8·6*
20	*37·4*	*31·8*	*28·1*	*25·7*	*19·8*	*16·1*	*11·4*
25	*46·8*	*39·8*	*35·1*	*32·1*	*24·7*	*20·1*	*14·3*

D / CV	2	5	10	15	20	25
2	*11*					
5		*11*	Less than 4			
10			*11*	*5*		
15			*16*	*11*	*7*	*5*
20	Greater than 31			*16*	*11*	7
25				*31*	*16*	*11*

Prepared by B. A. Ellem, CSIRO, Division of Mathematics and Statistics, Armidale, N.S.W., from Table 30, p. 250, of Pearson and Hartley (1972)

IX. REFERENCES

ALEXANDER, G. I. (Ed.) (1973). Manual of techniques for field investigations with beef cattle. Canberra, Australia: CSIRO.

ALEXANDER, R. H. & McGOWAN, M. (1966). The routine determination of *in vitro* digestibility of organic matter in forages—an investigation of the problems associated with continuous large-scale operation. *J. Br. Grassl. Soc.* **21,** 140-7.

ALLDEN, W. G. (1962). Rate of herbage intake of grazing time in relation to herbage availability. *Proc. Aust. Soc. Anim. Prod.* **4,** 163-6.

ALLDEN, W. G. (1968). Undernutrition of the Merino sheep and its sequelae. IV. Herbage consumption and utilizaton of feed for wool production following growth restrictions imposed at two stages of early post-natal life in a Mediterranean environment. *Aust. J. Agric. Res.* **19,** 997-1007.

ALLDEN, W. G. (1969). The summer nutrition of weaner sheep: the voluntary feed intake, body weight change, and wool production of sheep grazing the mature herbage of sown pasture in relation to the intake of dietary energy under a supplementary feeding regime. *Aust. J. Agric. Res.* **20,** 499-512.

ANDERSON, L. L. (1969). Research techniques in physiology of reproduction in domestic female mammals. *In* Techniques and procedures in animal science research. New York, Albany: American Society of Animal Science, pp. 111-29.

ARMSTRONG, D. G. (1964). Evaluation of artificially dried grass as a source of energy for sheep. II. The energy value of cocksfoot, timothy and two strains of rye-grass at varying stages of maturity. *J. Agric. Sci.* **62,** 399-416.

ARNOLD, G. W. & BUSH, I. G. (1963). Equipment for nutritional studies on grazing animals. *Aust. CSIRO Div. Plant Ind. Field Stn. Rec.* **2,** 59-62.

ARNOLD, G. W. & DUDZINSKI, M. L. (1963). The use of faecal nitrogen as index of estimating the consumption of herbage by grazing animals. *J. Agric. Sci.* **61,** 33-43.

ARNOLD, G. W. & DUDZINSKI, M. L. (1967). Comparison of faecal nitrogen regressions and *in vitro* estimates of diet digestibility for estimating the consumption of herbage by grazing animals. *J. Agric. Sci.* **68,** 213-9.

ARNOLD, G. W., McMANUS, W. R., BUSH, I. G. & BALL, J. (1964). The use of sheep fitted with oesophageal fistulas to measure diet quality. *Aust. J. Exp. Agric. Anim. Husb.* **4,** 71-9.

BAKER, A. A. (1965). Comparison of heat mount detectors and classical methods for detecting heat in beef cattle. *Aust. Vet. J.* **41,** 360-1.

BALCH, C. C., JOHNSON, V. W. & MACHIN, C. (1962). Housing and equipment for balance studies with cows. *J. Agric. Sci.* **59,** 355-8.

BALLINGER, C. E. & DUNLOP, A. A. (1946). An apparatus for the collection of faeces from the cow. *N.Z. J. Sci. Tech.* **27**(A), 509-20.

BARKER, A. S., CRAY, A. S., FOOT, A. S., IVINS, J. D., JONES, L. I. & WILLIAMS, T. E. (1955). The assessment and recording of the utilized output of grassland. *J. Br. Grassl. Soc.* **10,** 67-84.

BASSETT, E. G. (1952). Apparatus for continuous collection of urine from the grazing ewe. *N.Z. J. Sci. Tech.* **34**(A), 76-81.

BEESTON, J. W. U. & HOGAN, J. P. (1960). The estimation of pasture intake by the grazing sheep. *Proc. Aust. Soc. Anim. Prod.* **3,** 79-82.

BERG, R. T. & BUTTERFIELD, R. M. (1976). New concepts of cattle growth. Australia: Sydney University Press.

BLAXTER, K. L. (1961). The utilization of the energy of food by ruminants. *Eur. Assoc. Anim. Prod. Publ. No.* 10, 211-25.

BLAXTER, K. L. (1964). Utilization of the metabolizable energy of grass. *Proc. Nutr. Soc.* **23,** 62-70.

BLAXTER, K. L. (1967). The energy metabolism of ruminants. London: Hutchinson, 2nd edn.

BLAXTER, K. L. & CLAPPERTON, J. L. (1965). Prediction of the amount of methane produced by ruminants. *Br. J. Nutr.* **19,** 511-22.

BLAXTER, K. L., CLAPPERTON, J. L. & MARTIN, A. K. (1966). The heat of combustion of the urine of sheep and cattle in relation to its chemical composition and to diet. *Br. J. Nutr.* **20,** 449-60.

BLAXTER, K. L., CLAPPERTON, J. L. & WAINMAN, F. W. (1966). Utilization of the energy and protein of the same diet by cattle of different ages. *J. Agric. Sci.* **67,** 67-75.

BLAXTER, K. L., GRAHAM, N. MC. & WAINMAN, F. W. (1956). Some observations on the digestibility of food by sheep, and on related problems. *Br. J. Nutr.* **10,** 69-91.

BLAXTER, K. L. & MITCHELL, H. H. (1948). The factorization of the protein requirements of ruminants and of the protein values of feeds with particular reference to the significance of the metabolic faecal nitrogen. *J. Anim. Sci.* **7,** 351-72.

BLAXTER, K. L., WAINMAN, F. W. & DAVIDSON, J. L. (1966). The voluntary intake of food by sheep and cattle in relation to their energy requirements for maintenance. *Anim. Prod.* **8,** 75-83.

BLIGH, J. & HEAL, J. W. (1974). The use of radio-telemetry in the study of animal physiology. *Proc. Nutr. Soc.* **33,** 173-81.

BLUNTZER, J. S. & SIMS, P. L. (1976). Improved technique for estimating milk production from range cows. *J. Range Mgmt.* **29,** 168-69.

BOFINGER, V. J. & WHEELER, J. L. (Eds.) (1975). Developments in field experiment design and analysis. *Commonw. Bur. Pastures Field Crops, Hurley, Bull.* 50.

BORDER, J. R., HARRIS, L. E. & BUTCHER, J. E. (1963). Apparatus for obtaining sustained quantitative collections of urine from male cattle grazing pasture or range. *J. Anim. Sci.* **22,** 521-5.

BORRIE, J., REDSHAW, N. R. & DOBBINSON, T. L. (1973). Silastic tracheal bifurcation prosthesis with subterminal Dacron suture cuffs. *J. Thorac. Cardiovasc. Surg.* **65,** 956-62.

BRADEN, A. W. H. & BAKER, A. A. (1973). Reproduction in sheep and cattle. *In* The pastural industries of Australia. Eds. G. Alexander & O. B. Williams. Australia: Sydney University Press, pp. 269-302.

BRETT, D. J., CORBETT, J. L. & INSKIP, M. W. (1972). Estimation of the energy value of ewe milk. *Proc. Aust. Soc. Anim. Prod.* **9,** 286-91.

BRODY, S. (1945). Bioenergetics and growth. New York: Reinhold Publishing Co.

BRYANT, A. M., MURRAY, R. M. & LENG, R. A. (1973). Measurement of the turnover of methane in the rumen: a possible objective assessment of the degree of bloat. *In* Bloat. Eds. R. A. Leng & J. R. McWilliam. Australia, Armidale: University of New England, pp. 75-6.

BURTON, J. H. & REID, J. T. (1969). Interrelationships among energy input, body size, age and body composition of sheep. *J. Nutr.* **97,** 517-24.

BUTCHER, J. E. & HARRIS, L. E. (1957). Creatinine as an index material for evaluating ruminant nutrition. *J. Anim. Sci.* **16,** 1020.

BUTLER, G. W. & BAILEY, R. W. (Eds.) (1973). Chemistry and biochemistry of herbage (3 vols.) London: Academic Press.

BUTTERWORTH, M. H. (1964). The digestible energy content of some tropical forages. *J. Agric. Sci.* **64,** 319-21.

CAMPBELL, J. B. (1969). Experimental methods for evaluating herbage. *Can. Dept. Agric. Publ.* 1315.

CANAWAY, R. J., RAYMOND, W. F. & TAYLER, J. C. (1955). The automatic recording of animal behaviour in the field. *Electron. Eng.* **27,** 102-5.

CASTLE, M. E., FOOT, A. S. & HALLEY, R. J. (1950). Some observations on the behaviour of dairy cattle with particular reference to grazing. *J. Dairy Res.* **17,** 215-30.

CHAMBERS, A. R. M., MILNE, J. A., RUSSEL, A. J. F. & WHITE, I. R. (1975). An apparatus for the measurement and sampling of urine from grazing female sheep. *Proc. Nutr. Soc.* **34,** 71A.

CHAPMAN, H. W. (1964). Oesophageal fistulation and cannulation in sheep and cattle. *Aust. Vet. J.* **40,** 64-6.

CHAPMAN, H. W. & HAMILTON, F. J. (1962). Oesophageal fistulation of calves. *Aust.Vet.J.* **38,** 400.

CHAPMAN, R. E. (1960). Measurement of wool samples. *In* The biology of the fleece. *Aust. CSIRO Anim. Res. Lab. Tech. Pap. No.* 3, pp. 97-108.

CHAPMAN, R. E. & WHEELER, J. L. (1963). Dye-banding: A technique for fleece growth studies. *Aust. J. Sci.* **26,** 53-4.

CHARLES, D. D. (1964). Classifying trade beef by specifications. *Aust. Vet. J.* **40,** 27-9.

CHARLES, D. D., BUTTERFIELD, R. M. & FRANCIS, J. (1965). Marketing beef by specifications. *Agriculture, Lond.* **72,** 167-70.

CHRISTIAN, K. R. & COUP, M. R. (1954). Measurement of feed intake by grazing cattle and sheep. VI. The determination of chromic oxide in faeces. *N.Z. J. Sci. Tech.* **36**(A), 328-30.

CONWAY, A. G. (1973). Voluntary intake and stocking rate equivalents for grazing cattle. *Ir. J. Agric. Res.* **12,** 193-204.

COOK, C. W. (1964). Symposium on nutrition of forages and pastures: collecting forage samples representative of ingested material of grazing animals for nutritional studies. *J. Anim. Sci.* **23,** 265-70.

COOK, C. W. & HARRIS, L. E. (1951). A comparison of the lignin ratio technique and the chromogen method of determining digestibility and forage consumption of desert range plants by sheep. *J. Anim. Sci.* **10,** 565-73.

COOK, C. W., STODDART, L. A. & HARRIS L. E. (1952). Determining the digestibility and metabolizable energy of winter range plants by sheep. *J. Anim. Sci.* **11,** 578-90.

COOP, I. E. (1965). A review of the ewe equivalent system. *N.Z. Agric. Sci.* **1**(3), 13-8.

CORBETT, J. L. (1968). Variation in the yield and composition of milk of grazing Merino ewes. *Aust. J. Agric. Res.* **19,** 283-94.

CORBETT, J. L. (1969). The nutritional value of grassland herbage. *In* International encyclopedia of food and nutrition, Vol. 17, Part 2. Nutrition of animals of agricultural importance. Ed. D. Cuthbertson, Oxford: Pergamon Press, pp. 593-644.

CORBETT, J. L. & BOYNE, A. W. (1958). The effects of a low-protein food supplement on the yield and composition of milk from grazing dairy cows and on the composition of their diet. *J. Agric. Sci.* **51,** 95-107.

CORBETT, J. L., FARRELL, D. J., LENG, R. A., MCCLYMONT, G. L. & YOUNG, B. A. (1971). Determination of the energy expenditure of penned and grazing sheep from estimates of carbon dioxide entry rate. *Br. J. Nutr.* **26,** 277-91.

CORBETT, J. L., FURNIVAL, E. P., INSKIP, M. W., PEREZ, C. J. & PICKERING, F. S. (1976a). Nutrition and growth of lambs grazing lucerne or phalaris. *Proc. Aust. Soc. Anim. Prod.* **11,** 329-32.

CORBETT, J. L. & GREENHALGH, J. F. D. (1960). Measurement of the quantities of herbage consumed by grazing animals. *In* Chemical aspects of the production and use of grass.

Soc. Chem. Ind., London, Monogr. No. 9.

CORBETT, J. L., GREENHALGH, J. F. D., MCDONALD, I. & FLORENCE, E. (1960). Excretion of chromium sesquioxide administered as a component of paper to sheep. *Br. J. Nutr.* **14,** 289-99.

CORBETT, J. L., LENG, R. A. & YOUNG, B. A. (1969). Measurement of energy expenditure by grazing sheep and the amount of energy supplied by volatile fatty acids produced in the rumen. *Eur. Assoc. Anim. Prod. Publ. No.* 12, pp. 177-86.

CORBETT, J. L., LYNCH, J. J., NICOL, G. R. & BEESTON, J. W. U. (1976b). A versatile peristaltic pump designed for grazing lambs. *Lab. Pract.* **25,** 458-62.

CRABTREE, R. M., HOUSEMAN, R. A. & KAY, M. (1974). The estimation of body composition in beef cattle by deuterium oxide dilution. *Proc. Nutr. Soc.* **33,** 74A-5A.

CRAMPTON, E. W. (1956). Applied animal nutrition. The use of feedstuffs in the formulation of livestock rations. San Francisco: W. H. Freeman & Co.

CRESSWELL, E. & HARRIS, L. E. (1959). An improved rangemeter for sheep. *J. Anim. Sci.* **18,** 1447-51.

CROKER, B. H. (1959). A method of estimating the botanical composition of the diet of sheep. *N.Z. J. Agric. Res.* **2,** 72-85.

CUNNINGHAM, H. M., FREDERICK, G. L. & BRISSON, G. J. (1955). Application of an inflatable urethral catheter for urine collection from cows. *J. Dairy Sci.* **38,** 997-9.

DAVIDSON, J. (1954a). Procedure for the extraction, separation and estimation of the major fat-soluble pigments of hay. *J. Sci. Food Agric.* **5,** 1-7.

DAVIDSON, J. (1954b). A comparative study of the pigments in microbial fractions from the sheep's rumen and in the corresponding diet. *J. Sci. Food Agric.* **5,** 87-92.

DAVIDSON, J. (1954c). The chromogen method for determining the digestibility of dried grass by sheep. *J. Sci. Food Agric.* **5,** 209-12.

DAY, N., JENKINSON, D. MCE. & WALKER-LOVE, J. (1971). Freeze-branding of young beef animals. *Anim. Prod.* **13,** 93-9.

DEIJS, W. B. & BOSMAN, M. S. M. (1955). Plant pigments in digestion trial studies. *Recl. Trav. Chim. Pays-Bas* **74,** 1207-16.

DONEY, J. M. (1966). Breed differences in response of wool growth to annual nutritional and climatic cycles. *J. Agric. Sci.* **67,** 25-30.

DONNELLY, J. R. & FREER, M. (1974). Prediction of body composition in live sheep. *Aust. J. Agric. Res.* **25,** 825-34.

DOWNES, A. M. & HUTCHINSON, J. C. D. (1969). Effect of low skin surface temperature on wool growth. *J. Agric. Sci.* **72,** 155-8.

DOWNES, A. M. & MCDONALD, I. W. (1964). The chromium-51 complex of ethylenediamine tetraacetic acid as a soluble rumen marker. *Br. J. Nutr.* **18,** 153-62.

DOWNES, A. M. & SHARRY, L. F. (1971). Measurement of wool growth and its response to nutritional changes. *Aust. J. Biol. Sci.* **24,** 117-30.

DREW, K. R. & KELLY, R. W. (1975). Handling deer, run in confined areas. *Proc. N.Z. Soc. Anim. Prod.* **35,** 213-8.

DUDZIŃSKI, M. L. & ARNOLD, G. W. (1967). Aerial photography and statistical analysis for studying behaviour patterns of grazing animals. *J. Range Mgmt.* **20,** 77-83.

DUDZIŃSKI, M. L. & ARNOLD, G. W. (1973). Comparisons of diets of sheep and cattle grazing together on sown pastures on the Southern Tablelands of New South Wales by principal components analysis. *Aust. J. Agric. Res.* **24,** 899-912.

DUDZIŃSKI, M. L., PAHL, P. J. & ARNOLD, G. W. (1969). Quantitative assessment of grazing behaviour of sheep in arid areas. *J. Range Mgmt.* **22,** 230-5.

DÝRMUNDSSON, Ó. R. (1973a). Puberty and early reproductive performance in sheep. I. Ewe lambs. *Anim. Breed. Abstr.* **41,** 273-89.

DÝRMUNDSSON, Ó. R. (1973b). Puberty and early reproductive performance in sheep. II. Ram lambs. *Anim. Breed Abstr.* **41,** 419-30.

EGAN, A. R. (1965). Nutritional status and intake regulation in sheep. III. The relationship between improvement of nitrogen status and increase in voluntary intake of low-protein roughages by sheep. *Aust. J. Agric. Res.* **16,** 463-72.

ENGELS, E. A. N. & HUGO, J. M. (1969). A technique for the collection of urine of grazing Merino wethers. *Proc. S. Afr. Soc. Anim. Prod.,* **8,** 209.

ENGELS, E. A. N., INSKIP, M. W. & CORBETT, J. L. (1976). Effect of change in respiratory quotient on the relationship between carbon dioxide entry rate in sheep and their energy expenditure. *Eur. Ass. Anim. Prod. Publ.* **19,** 339-42.

ERIZIAN, E. (1932). Eine neue Methode zur Bestimmung der vom Vieh gefressenen Menge Weidefutters. *Z. Zûcht.* **25,** 443-59.

EWBANK, R. (1967). Nursing and suckling behaviour amongst Clun Forest ewes and lambs. *Anim. Behav.* **15,** 251-8.

FALVEY, L. & WOOLLEY, A. (1974). Losses from cattle faeces during chemical analysis. *Aust. J. Exp. Agric. Anim. Husb.* **14,** 716-9.

FARRELL, D. J., CORBETT, J. L. & LENG, R. A. (1970). Automatic sampling of blood and ruminal fluid of grazing sheep. *Res. Vet. Sci.* **11,** 217-20.

FARRELL, D. J., LENG, R. A. & CORBETT, J. L. (1972a). Undernutrition in grazing sheep. I. Changes in the composition of the body, blood and rumen contents. *Aust. J. Agric. Res.* **23,** 483-97.

FARRELL, D. J., LENG, R. A. & CORBETT, J. L. (1972b). Undernutrition in grazing sheep. II. Calorimetric measurements on sheep taken from pasture. *Aust. J. Agric. Res.* **23,** 499-509.

FISHER, A. V. (1975). The accuracy of some body measurements on live beef steers. *Livestock Prod. Sci.* **2,** 357-66.

FLATT, W. P., WALDO, D. R., SYKES, J. F. & MOORE, L. A. (1958). A proposed method of indirect calorimetry for energy metabolism studies with large animals under field conditions. *Eur. Assoc. Anim. Prod. Publ. No.* 8, pp. 101-9.

FOOTE. R. H. (1969). Research techniques to study reproductive physiology in the male. *In* Techniques and procedures in animal science research. New York, Albany: American Society of Animal Science, pp. 80-110.

FORBES, R. M. (1949). Some difficulties involved in the use of feacal nitrogen as a measure of dry matter intake of grazing animals. *J. Anim. Sci.* **8,** 19-23.

FREE, J. C., HANSEN, R. M. & SIMS, P. L. (1970). Estimating dryweights of foodplants in faeces of herbivores. *J. Range Mgmt.* **23,** 300-302.

GAILLARD, B. D. E. (1966). Calculation of the digestibility for ruminants of roughages from the contents of cell-wall constituents. *Neth. J. Agric. Sci.* **14,** 215-23.

GAILLARD, D. E. & NIJKAMP, H. J. (1968). Calculation of the digestibility for ruminants of roughages from their contents of cell-wall constituents. II. Time-saving method of analysis. *Neth. J. Agric. Sci.* **16,** 21-4.

GALLUP, W. D. & BRIGGS, H. M. (1948). The apparent digestibility of prairie hay of variable protein content with some observations of faecal nitrogen excretion by steers in relation to their dry matter intake. *J. Anim. Sci.* **7,** 110-8.

GARRETT, W. N. & HINMAN, N. (1969). Re-evaluation of the relationship between carcass density and body composition of beef steers. *J. Anim. Sci.* **28,** 1-5.

GARTNER, R. J. W., CALLOW, L. L., GRANZIEN, C. K. & PEPPER, P. M. (1969). Variation in the concentration of blood constituents in relation to the handling of cattle. *Res. Vet. Sci.* **10,** 7-12.

GARTNER, R. J. W., DIMMOCK, C. K., STOKOE, J. & LAWS, L. (1970). Effect of frequency of handling sheep on blood constituents, with special reference to potassium and sodium and the repeatability of the estimates. *Queensl. J. Agric. Anim. Sci.* **27,** 405-10.

GESSAMAN, J. A. (Ed.) (1973). Ecological energetics of homeotherms. *Utah State Univ. Press Monogr. Ser. No.* 20.

GIFFORD, W. (1953). Records-of-performance tests for beef cattle in breeding herds; milk production of dams and growth of calves. *Bull. Ark. Agric. Exp. Sta.,* No. 531.

GLOVER, J., DUTHIE, D. W. & FRENCH, M. H. (1957). The apparent digestibility of crude protein by the ruminant. I. A synthesis of the results of digestibility trials with herbage and mixed feeds. *J. Agric. Sci.* **48,** 373-8.

GOLLEY, F. B. & BUECHNER, H. K. (Eds.) (1968). A practical guide to the study of the productivity of large herbivores. Int. Biol. Program. Handb. No. 7, Oxford: Blackwell Sci. Publ.

GORDON, I. (1975). Hormonal control of reproduction in sheep. *Proc. Br. Soc. Anim. Prod.* **4,** (New ser.), 79-93.

GRAHAM, N. McC. (1962). Measurement of the heat production of sheep: The influence of training and of a tranquillizing drug. *Proc. Aust. Soc. Anim. Prod.* **4,** 138-44.

GRAHAM, N. McC. (1969). Relation between metabolizable and digestible energy in sheep and cattle. *Aust. J. Agric. Res.* **20,** 1117-22.

GRAHAM, N. McC. & SEARLE, T. W. (1972). Balances of energy and matter in growing sheep at several ages, body weights, and planes of nutrition. *Aust. J. Agric. Res.* **23,** 97-108.

GRASSLAND RESEARCH INSTITUTE, HURLEY (1961). Research techniques in use at the Grassland Research Institute, Hurley. *Commonw. Bur. Pastures Field Crops, Hurley. Berkshire, Bull.* 45.

GREENHALGH, J. F. D. & CORBETT, J. L. (1960). The indirect estimation of the digestibility of pasture herbage. I. Nitrogen and chromogen as faecal index substances. *J. Agric. Sci.* **55,** 371-6.

GREENHALGH, J. F. D., CORBETT, J. L. & MCDONALD, I. (1960). The indirect estimation of the digestibility of pasture herbage. II. Regressions of digestibility on faecal nitrogen concentration; their determination in continuous digestibility trials and the effect of various factors on their accuracy. *J. Agric. Sci.* **55,** 377-86.

GREENHALGH, J. F. D., REID, G. W. & MCDONALD, I. (1966). The indirect estimation of the digestibility of pasture herbage. IV. Regressions of digestibility on faecal nitrogen concentration: effects of different fractions of the herbage upon within- and between-period regressions. *J. Agric. Sci.* **66,** 277-83.

GRIFFITHS, M., BARKER, R. & MACLEAN, L. (1974). Further observations on the plants eaten by kangaroos and sheep grazing together in a paddock in south-western Queensland. *Aust. Wildl. Res.* **1,** 27-43.

GRIMES, R. C., WATKIN, B. R. & MAY, P. F. (1965). The botanical and chemical analysis of herbage samples obtained from sheep fitted with oesophageal fistulae. *J. Br. Grassl. Soc.* **20,** 168-73.

HAFEZ, E. S. E. (Ed.) (1969). The behaviour of domestic animals. Baltimore, M. D.: Williams & Wilkins, 2nd Edn.

HALL, D. G. & HAMILTON, B. A. (1975). Estimation of the botanical composition of oesophageal extrusa samples. 2. A comparison of manual separation and a microscope point technique. *J. Br. Grassl. Soc.* **30,** 273-7.

HAMILTON, B. A. & HALL, D. G. (1975). Estimation of the botanical composition of oesophageal extrusa samples. 1. A modified microscope point technique. *J. Br. Grassl. Soc.* **30,** 229-35.

HANCOCK, J. (1950). Grazing habits of dairy cows in New Zealand. *Emp. J. Exp. Agric.* **18,** 249-63.

HANCOCK, J. (1954a). Studies of grazing behaviour in relation to grassland management. I. Variations in grazing habits of dairy cattle. *J. Agric. Sci.* **44,** 420-33.

HANCOCK, J. (1954b). Studies in grazing behaviour of dairy cattle. II. Bloat in relation to grazing behaviour. *J. Agric. Sci.* **45,** 80-95.

HANCOCK, J. & MCMEEKAN, C. P. (1954). Studies of grazing behaviour in relation to grassland management. III. Rotational compared with continuous grazing. *J. Agric. Sci.* **45,** 96-103.

HARKER, K. W., TORELL, D. T. & VAN DYNE, G. M. (1964). Botanical examination of forage from esophageal fistulas in cattle. *J. Anim. Sci.* **23,** 465-9.

HARRIS, L. E. (1970). Nutritional research techniques for domestic and wild animals. Vol. I. An international record system and procedures for analyzing samples. Utah State Univ. USA, Publ. by L. E. Harris.

HEADY, H. F. & TORELL, D. T. (1959). Forage preference exhibited by sheep with esophageal fistulas. *J. Range Mgmt.* **12,** 28-34.

HEANEY, D. P. & PIGDEN, W. J. (1963). Interrelationships and conversion factors between expressions of the digestible energy value of forages. *J. Anim. Sci.* **22,** 956-60.

HECKER, J. F. (1974). Experimental surgery on small ruminants. London: Butterworths.

HERCUS, B. H. (1960). Plant cuticle as an aid to determining the diet of grazing animals. *Proc. VIIIth Grassl. Congr.,* pp. 443-7.

HOEHNE, O. E., CLANTON, D.C. & STREETER, C. L. (1967). Chemical changes in esophageal fistula samples caused by salivary contamination and sample preparation. *J. Anim. Sci.* **26,** 628-31.

HOFMAN, R. R. & STEWART, D. R. M. (1972). Grazer or browser: A classification based on the stomach-structure and feeding habits of East African ruminants. *Mammalia,* **36,** 226-40.

HOGAN, J. P. & WESTON, R. H. (1967). The digestion of chopped and ground roughages by sheep. II. The digestion of nitrogen and some carbohydrate fractions in the stomach and intestines. *Aust. J. Agric. Res.* **18,** 803-19.

HOLDER, J. M. (1962). Supplementary feeding of grazing sheep; its effect on pasture intake. *Proc. Aust. Soc. Anim. Prod.* **4,** 154-9.

HOLLEMAN, D. F., WHITE, R. G. & LUICK, J. R. (1975). New isotope methods for estimating milk intake and yield. *J. Dairy Sci.* **58,** 1814-21.

HOLMES, W. & JONES, J. G. W. (1964). The efficiency of utilization of fresh grass. *Proc. Nutr. Soc.* **23,** 88-99.

HOLTER, J. A. & REID, J. T. (1959). Relationship between the concentrations of crude protein and apparently digestible protein in forages. *J. Anim. Sci.* **18,** 1339-49.

HUGHES, G. P. & HARKER, K. W. (1950). The technique of weighing bullocks on summer grass. *J. Agric. Sci.* **40,** 403-9.

HUGHES, J. G. (1976). Short-term variation in animal live weight and reduction of its effect on weighing. *Anim. Breed. Abstr.* **44,** 111-8.

HUGHES, J. W. (1963). Equipment for the separate and total collection of faeces and urine from dairy cattle. *N.Z. J. Agric. Res.* **6,** 127-39.

HUTCHINSON, K. J. (1971). Productivity and energy flow in grazing/fodder conservation systems. *Herb. Abstr.* **41,** 1-10.

HUTTON, J. B. (1961). Studies of the nutritive value of New Zealand dairy pastures. I. Seasonal changes in some chemical components of pastures. *N.Z. J. Agric. Res.* **4,** 583-90.

HUTTON, J. B. (1962). Studies of the nutritive value of New Zealand dairy pastures.

II. Herbage intake and digestibility studies with dry cattle. *N.Z. J. Agric. Res.* **5,** 409-24.

JEFFERY, H. J. (1970). The length of change-over periods in change-over design with grazing cattle. *Aust. J. Exp. Agric. Anim. Husb.* **10,** 691-3.

JEFFERY, H. J. (1971). Assessment of faecal nitrogen as an index for estimating digestibility and intake of food by sheep on *Pennisetum clandestinum* based pastures. *Aust. J. Exp. Agric. Anim. Husb.* **11,** 393-6.

JOHNSON, E. R. & CHARLES, D. D. (1976). An evaluation of the Australian beef carcass appraisal system. *Aust. Vet. J.* **52,** 149-54.

JOHNSON, R. R. (1969). Techniques and procedures for *in vitro* and *in vivo* rumen studies. *In* Techniques and procedures in animal science research. New York, Albany: American Society of Animal Science, pp. 175-96.

JONES, D. I. H. & HAYWARD, M. V. (1975). The effect of pepsin pretreatment of herbage on the prediction of dry matter digestibility from solubility in fungal cellulose solutions. *J. Sci. Food Agric.* **26,** 711-8.

JONES, R. J. & COWPER, L. J. (1975). A lightweight electronic device for measurement of grazing time of cattle. *Trop. Grassl.* **9,** 235-41.

KANE, E. A., JACOBSON, W. C. & DAMEWOOD, P. M. (1959). Use of radioactive chromium oxide in digestibility determinations. *J. Dairy Sci.* **42,** 1359-66.

KAVANAGH, D. & ZANDER, H. (1965). A versatile recording system for studies of mastication. *Med. Electron. Biol. Eng.* **3,** 291-300.

KENNEDY, W. K., CARTER, A. H. & LANCASTER, R. J. (1959). Comparison of faecal pigments and faecal nitrogen as digestibility indicators in grazing cattle studies. *N.Z. J. Agric. Res.* **2,** 627-38.

KILGOUR, R. (1974). Potential value of animal behaviour studies in animal production. *Proc. Aust. Soc. Anim. Prod.* **10,** 286-98.

KILGOUR, R., PEARSON, A. J. & DE LANGEN, H. (1975). Sheep dispersal patterns on hill country: techniques for study and analysis. *Proc. N.Z. Soc. Anim. Prod.* **35,** 191-7.

KOTB, A. R. & LUCKEY, T. D. (1972). Markers in nutrition. *Nutr. Abstr. Rev.* **42,** 813-45.

KYDD, W. L. & MULLINS, G. (1963). A telemetry system for intraoral pressures. *Archs. Oral Biol.* **8,** 253-6.

LAMBOURNE, L. J. & REARDON, T. F. (1962). Use of ' seasonal ' regressions in measuring feed intake of grazing animals. *Nature, U.K.,* **196,** 961-2.

LAMBOURNE, L. J. & REARDON, T. F. (1963a). The use of chromium oxide to estimate the faecal output of Merinos. *Aust. J. Agric. Res.* **14,** 239-56.

LAMBOURNE, L. J. & REARDON, T. F. (1963b). Effect of environment on the maintenance requirements of Merino wethers. *Aust. J. Agric. Res.* **14,** 272-93.

LAMMING, G. E., HAFS, H. D. & MANNS, J. A. (1975). Hormonal control of reproduction in cattle. *Proc. Br. Soc. Anim. Prod.* **4** (New ser.), 71-8.

LAMOND, D. R., HOLMES, J. H. G. & HAYDOCK, K. P. (1969). Estimation of yield and composition of milk produced by grazing beef cows. *J. Anim. Sci.* **29,** 606-11.

LANCASTER, R. J. (1949). Estimation of digestibility of grazed pasture from faeces nitrogen. *Nature, UK,* **163,** 330-1.

LANCASTER, R. J. & BARTRUM, M. P. (1954). Measurement of feed intake by grazing cattle and sheep. IV. A source of error in the chromogen technique of estimating the digestibility of fodders. *N.Z. J. Sci. Tech.* **35** (A), 489-96.

LANG, D. R. & HIGHT, G. K. (1969). Using the bull mating harness. *N.Z. J. Agric.* **119**(2), 89-91.

LANG, D. R., HIGHT, G. K., ULJEE, A. E. & YOUNG, J. (1968). A marking device for detecting oestrous activity of cattle. *N.Z. J. Agric. Res.* **11,** 955-8.

LANGLANDS, J. P. (1966a). Creatinine as an index substance for estimating the urinary excretion of nitrogen and potassium by grazing sheep. *Aust. J. Agric. Res.* **17,** 757-63.

LANGLANDS, J. P. (1966b). Studies on the nutritive value of the diet selected by grazing sheep. I. Differences in composition between herbage consumed and material collected from oesophageal fistulae. *Anim. Prod.* **8,** 253-9.

LANGLANDS, J. P. (1967a). Studies on the nutritive value of the diet selected by grazing sheep. II. Some sources of error when sampling oesophageally fistulated sheep at pasture. *Anim. Prod.* **9,** 167-75.

LANGLANDS, J. P. (1967b). Studies on the nutritive value of the diet selected by grazing sheep. III. A comparison of oesophageal fistula and faecal index techniques for the indirect estimation of digestibility. *Anim. Prod.* **9,** 325-31.

LANGLANDS, J. P. (1969a). Studies on the nutritive value of the diet selected by grazing sheep. IV. Variation in the diet selected by sheep differing in age, breed, sex, strain and previous history. *Anim. Prod.* **11,** 369-78.

LANGLANDS, J. P. (1969b). Studies on the nutritive value of the diet selected by grazing sheep. V. Further studies of the relationship between digestibility estimated *in vitro* from oesophageal fistula samples and from faecal and dietary composition. *Anim. Prod.* **11,** 379-87.

LANGLANDS, J. P. (1969c). The feed intake of sheep supplemented with varying quantities of wheat while grazing pastures differing in herbage availability. *Aust. J. Agric. Res.* **20,** 919-24.

LANGLANDS, J. P. (1972). Growth and herbage consumption of grazing Merino and Border Leicester lambs reared by their mothers or fostered by ewes of the other breed. *Anim. Prod.* **14,** 317-22.

LANGLANDS, J. P. (1973). Milk and herbage intakes by grazing lambs born to Merino ewes and sired by Merino, Border Leicester, Corriedale, Dorset Horn and Southdown rams. *Anim. Prod.* **16,** 285-91.

LANGLANDS, J. P. (1974). Studies on the nutritive value of the diet selected by grazing sheep. VII. A note on hand plucking as a technique for estimating dietary composition. *Anim. Prod.* **19,** 249-52.

LANGLANDS, J. P. (1975). Techniques for estimating nutrient intake and its utilization by the grazing ruminant. *In* Digestion and metabolism in the ruminant; Proc. IVth Int. Symp. Ruminant Physiol. Eds. I. W. McDonald & A. C. I. Warner. University of New England Publications Unit, pp. 320-32.

LANGLANDS, J. P. & BENNETT, I. L. (1973a). Stocking intensity and pastoral production. II. Herbage intake of Merino sheep grazed at different stocking rates. *J. Agric. Sci.* **81,** 205-9.

LANGLANDS, J. P. & BENNETT, I. L. (1973b). Stocking intensity and pastoral production. III. Wool production, fleece characteristics, and the utilization of nutrients for maintenance and wool growth by Merino sheep grazed at different stocking rates. *J. Agric. Sci.* **81,** 211-8.

LANGLANDS, J. P. & BOWLES, J. E. (1973). Studies on the nutritive value of the diet selected by grazing sheep. VI. The use of tritiated water as a marker to estimate the composition of the diet from material collected by oesophageally fistulated sheep. *Anim. Prod.* **16,** 59-65.

LANGLANDS, J. P. & BOWLES, J. E. (1974). Herbage intake and production of Merino sheep grazing native and improved pastures at different stocking rates. *Aust. J. Exp. Agric. Anim. Husb.* **14,** 307-15.

LANGLANDS, J. P., CORBETT, J. L. & MCDONALD, I. (1963). The indirect estimation of the digestibility of pasture herbage. III. Regressions of digestibility on faecal nitrogen concentration: effects of species and individuality of animal and of the method of determining digestibility upon the relationship. *J. Agric. Sci.* **61,** 221-6.

LANGLANDS, J. P., CORBETT, J. L., McDONALD, I. & REID, G. W. (1963a). Estimation of the faeces output of grazing animals from the concentration of chromium sesquioxide in a sample of faeces. I. Comparison of estimates from samples taken at fixed times of day with faeces outputs measured directly. *Br. J. Nutr.* **17,** 211-8.

LANGLANDS, J. P., CORBETT, J. L., McDONALD, I. & REID, G. W. (1963b). Estimation of the faeces output of grazing animals from the concentration of chromium sesquioxide in a sample of faeces. 2. Comparison of estimates from samples taken at fixed times of day with estimates from samples collected from the sward. *Br. J. Nutr.* **17,** 219-26.

LANGLANDS, J. P., CORBETT, J. L., McDONALD, I. & REID, G. W. (1963c). Estimates of the energy required for maintenance by adult sheep. 2. Grazing sheep. *Anim. Prod.* **5,** 11-6.

LANGLANDS, J. P. & HOLMES, C. R. (1975). Consumption of pluronics administered in drinking water and roller drums to grazing beef cattle. *Aust. J. Exp. Agric. Anim. Husb.* **15,** 5-11.

LANGLANDS, J. P. & WHEELER, J. L. (1968). The dyebanding and tattooed patch procedures for estimating wool production and obtaining samples for the measurement of fibre diameter. *Aust. J. Exp. Agric. Anim. Husb.* **8,** 265-9.

LEDGER, H. P., GILLIVER, B. & ROBB, J. M. (1973). An examination of sample joint dissection and specific gravity techniques for assessing the carcass composition of steers slaughtered in commercial abattoirs. *J. Agric. Sci.* **80,** 381-92.

LENG, R. A. (1970). Formation and production of volatile fatty acids in the rumen. *In* Physiology of digestion and metabolism in the ruminant, Proc. IIIrd International Symposium. Ed. A. T. Phillipson, Newcastle-on-Tyne: Oriel Press Ltd., pp. 406-21.

LENG, R. A., MURRAY, R. M., NOLAN, J. V. & NORTON, B. W. (1975). Estimation of the intake of supplements by individual grazing ruminants. *Proc. IIIrd World Conference on Animal Production.* Sydney: University Press, pp. 587-91.

LESPERANCE, A. L. & BOHMAN, V. R. (1961). Apparatus for collecting excreta from grazing cattle. *J. Anim. Sci.* **20,** 503-5.

LIFSON, N., GORDON, G. B. & McCLINTOCK, R. M. (1955). Measurement of total carbon dioxide production by means of $D_2{}^{18}O$. *J. Appl. Physiol.* **7,** 704-10.

LITTLE, D. A. (1972). Studies on cattle with oesophageal fistulae. The relation of the chemical composition of feed to that of the extruded bolus. *Aust. J. Exp. Agric. Anim. Husb.* **12,** 126-30.

LITTLE, D. A. (1975). Studies on cattle with oesophageal fistulae: comparison of concentrations of mineral nutrients in feeds and associated boluses. *Aust. J. Exp. Agric. Anim. Husb.* **15,** 437-9.

LITTLE, D. A. & MORRIS, J. G. (1972). Prediction of the body composition of live cattle. *J. Agric. Sci.* **78,** 505-8.

LITTLE, D. A. & TAKKEN, A. (1970). Preparation of oesophageal fistulae in cattle under local anaesthesia. *Aust. Vet. J.* **46,** 335-7.

LOFGREEN, G. P. (1965). A comparative slaughter technique for determining net energy values with beef cattle. *Eur. Assoc. Anim. Prod. Publ. No.* 11, pp. 309-17.

LOGAN, V. S. & PIGDEN, W. J. (1969). Estimating herbage yield from energy intake of grazing ruminants. *In* Experimental methods for evaluating herbage. Ed. J. B. Campbell. Publ. Can. Dep. Agric., No. 1315. Ottowa: Queen's Printer for Can., pp. 140-51.

LOOSLI, J. K. & McDONALD, I. W. (1968). Nonprotein nitrogen in the nutrition of ruminants. *FAO Agric. Stud.* No. 75.

LYNCH, J. J. (1974). Behaviour of sheep and cattle in the more arid areas of Australia.

In Studies of the Australian arid zone. II. Animal production. Ed. A. D. Wilson. Australia, Melbourne: CSIRO Div. Land Res. Mgmt., pp. 37-49.

LYNCH, J. J. & ALEXANDER, G. (1973). Animal behaviour and the pastoral industries. *In* The pastoral industries of Australia. Eds. G. Alexander & O. B. Williams, Sydney: University Press, pp. 371-400.

McDONALD, I. W. (1968). The nutrition of grazing ruminants. *Nutr. Abstr. Rev.* **38,** 381-400.

McDONALD, P. & PURVES, D. (1957). The estimation of feed intake by sheep on a silage diet. *J. Br. Grassl. Soc.* **12,** 22-9.

MACFARLANE, W.V., HOWARD, B. & SIEBERT, B. D. (1969). Tritiated water in the measurement of milk intake and tissue growth of ruminants in the field. *Nature, UK* **221,** 578-9.

MACKAY, R. S. (1970). Bio-medical telemetry. New York: John Wiley & Sons, Inc. 2nd Edn.

McMANUS, W. R. (1962). Oesophageal fistulation studies in the sheep. *Aust. Vet. J.* **38,** 85-91.

McMANUS, W. R., ARNOLD, G. W. & HAMILTON, F. J. (1962). Improved techniques in oesophageal fistulation of sheep. *Aust. Vet. J.* **38,** 275-81.

McMANUS, W. R., DUDZINSKI, M. L. & ARNOLD, G. W. (1967). Estimation of herbage intake from nitrogen, copper, magnesium and silicon concentrations in faeces. *J. Agric. Sci.* **69,** 263-8.

McQUEEN, R. & VAN SOEST, P. J. (1975). Fungal cellulase and hemicellulase prediction of forage digestibility. *J. Dairy Sci.* **58,** 1482-91.

MACRAE, J. C. (1974). The use of intestinal markers to measure digestive function in ruminants. *Proc. Nutr. Soc.* **33,** 147-54.

MACRAE, J. C., CAMPBELL, D. R. & EADIE, J. (1975). Changes in the biochemical composition of herbage upon freezing and thawing. *J. Agric. Sci.* **84,** 125-31.

MACRAE, J. C. & EVANS, C. C. (1974). The use of inert ruthenium-phenanthroline as a digesta particulate marker in sheep. *Proc. Nutr. Soc.* **3,** 10A-11A.

MARSHALL. B., TORELL, D. T. & BREDON, R. M. (1967). Comparison of tropical forages of known composition with samples of these forages collected by oesophageal fistulated animals. *J. Range Mgmt.* **20,** 310-3.

MARTIN, A. K., MILNE, J. A. & MOBERLEY, P. (1975). Urinary quinol and orcinol outputs as indices of voluntary intake of heather. (*Calluna vulgaris* L. (Hull)) by sheep. *Proc. Nutr. Soc.* **34,** 70A.

METEOROLOGICAL OFFICE GREAT BRITAIN (1956). Handbook of meteorological instruments. I. Instruments for surface observations. London: H.M.S.O.

MICHELL, P. J. (1974). Gross energy levels in regrowths of six pasture species, and relations with digestibility and chemical composition. *Aust. J. Exp. Agric. Anim. Husb.* **14,** 33-7.

MILFORD, R. (1957). The value of faecal nitrogen and faecal crude fibre in estimating intake of four subtropical grass species. *Aust. J. Agric. Res.* **8,** 359-70.

MILFORD, R. & HAYDOCK, K. P. (1965). The nutritive value of protein in subtropical pasture species grown in south-east Queensland. *Aust. J. Exp. Agric. Anim. Husb.* **5,** 13-7.

MILFORD, R. & MINSON, D. J. (1965). The relation between the crude protein content and digestible crude protein content of tropical pasture plants. *J. Br. Grassl. Soc.* **20,** 177-9.

MINSON, D. J. & KEMP, C. D. (1961). Studies in the digestibility of herbage. IX. Herbage and faecal nitrogen as indicators of herbage organic matter digestibility. *J. Br. Grassl. Soc.* **16,** 76-9.

MINSON, D. J. & MCLEOD, M. N. (1970). The digestibility of temperate and tropical grasses. *Proc.XIth Int. Grassl. Congr.*, pp. 719-22.

MINSON, D. J. & MCLEOD, M. N. (1972). The *in vitro* technique: its modification for estimating digestibility of large numbers of tropical pasture samples. *Aust. CSIRO Div. Trop. Past. Tech. pap. No.* 8.

MINSON, D. J. & MILFORD, R. (1966). The energy values and nutritive value indices of *Digitaria decumbens, Sorghum almum* and *Phaseolus atropurpureus. Aust. J. Agric. Res.* **17,** 411-23.

MINSON, D. J. & MILFORD, R. (1967). *In vitro* and faecal nitrogen techniques for predicting the voluntary intake of *Chloris gayana. J. Br. Grassl. Soc.* **22,** 170-5.

MINSON, D. J. & MILFORD, R. (1968). Equipment and housing for intake and digestibility studies with large numbers of wethers. *J. Agric. Sci.* **71,** 381-2.

MINSON, D. J., TAYLOR, J. C., ALDER, F. E., RAYMOND, W. F., RUDMAN, J. E., LINE, C. & HEAD, M J. (1960). A method for identifying the faeces produced by individual cattle or groups of cattle grazing together. *J. Br. Grassl. Soc.* **15,** 86-8.

MITCHELL, H. H. (1962). Comparative nutrition of man and domestic animals. Vol. 1. New York: Academic Press.

MOIR, R. J. (1961). A note on the relationship between the digestible dry matter and the digestible energy content of ruminant diets. *Aust. J. Exp. Agric. Anim. Husb.* **1,** 24-6.

MOORE, R. W. (1967). A comparison of methods of estimating milk intake of lambs and milk yield of ewes at pasture. *Aust.J. Exp. Agric. Anim. Husb.* **7,** 137-40.

MORLEY, F. H. W. & SPEDDING, C. R. W. (1968). Agricultural systems and grazing experiments. *Herb. Abstr.* **38,** 279-87.

MOULE, G. R. (Ed.) (1965). Field investigation with sheep: a manual of techniques. Melbourne; CSIRO Publ.

MOULE, G. R., BRADEN, A. W. H. & LAMOND, D. R. (1963). The significance of oestrogens in pasture plants in relation to animal production. *Anim. Breed. Abstr.* **31,** 139-57.

NEWMAN, D. M. R. & MCLEOD, M. N. (1973). Accuracy of predicting digestibility of browse species using the *in vitro* technique. *J. Aust. Inst. Agric. Sci.* **39,** 67-8.

NICHOLS, G. de la M. (1966). Radio transmission of sheep's jaw movements. *N.Z.J. Agric Res.* **9,** 468-73.

NICOL, A. M. & IRVINE, C. H. G. (1973). Measurement of the milk consumption of suckling beef calves by an isotope dilution method. *Proc. N.Z. Soc. Anim. Prod.* **33** 176-83.

NICOL, A. M. & SHARAFELDIN, M. A. (1975). Observations on the behaviour of single-suckled calves from birth to 120 days. *Proc. N.Z. Soc. Anim. Prod.* **35,** 221-30.

NOLAN, J. V., BALL, F. M., MURRAY, R. M., NORTON, B. W. & LENG, R. A. (1974). Evaluation of a urea-molasses supplement for grazing cattle. *Proc. Aust. Soc. Anim. Prod.* **10,** 91-4.

NOLAN, J. V., NORTON, B. W., MURRAY, R. M., BALL, F. M., ROSEBY, F. B., ROHAN-JONES, W., HILL, M. K. & LENG, R. A. (1975). Body weight and wool production in grazing sheep given access to a supplement of urea and molasses: intake of supplement/response relationships. *J. Agric. Sci.* **84,** 39-48.

OBIOHA, F. C., CLANTON, D. C., RITTENHOUSE, L. R. & STREETER, C. L. (1970). Sources of variation in chemical composition of forage ingested by esophageal fistulated cattle. *J. Range Mgmt.* **23,** 133-6.

OLUBAJO, F. O. & OYENUGA, V. A. (1970). Digestibility of tropical pasture mixtures using the indicator technique. *J. Agric. Sci.* **75,** 175-81.

OSBOURN, D. F., TERRY, R. A., OUTEN, G. E., CAMMELL, S. B. & LANSLEY, P. R. (1971). Chemical and *in vitro* digestion procedures for the prediction of the digestibility of

forage crops by sheep. *Proc. Nutr. Soc.* **30,** 85A-6A.
OWEN, J. B. (1957). A study of the lactation and growth of hill sheep in their native environment and under lowland conditions. *J. Agric. Sci.* **48,** 387-411.
OWEN, J. B. (Ed.) (1974). The detection and control of breeding activity in farm animals Proc. Symp. Sch. Agric., Univ. Aberdeen. Aberdeen: Waverley Press.
PALADINES, O. L., REID, J. T., VAN NIEKERK, B. D. H. & BENSADOUN, A. (1964a). Relationship between the nitrogen content and the heat of combustion value of sheep urine. *J. Anim. Science.* **23,** 528-32.
PALADINES, O. L., REID, J. T., BENSADOUN, A. & VAN NIEKERK, B. D. H. (1964b). Heat of combustion values of the protein and fat in the body and wool of sheep. *J. Nutr.* **82,** 145-9.
PATIL, B. D., JONES, D. I. H. & HUGHES, R. (1969). Wool characteristics as an indication of nutritive attributes in herbage varieties. *Nature, UK,* **223,** 1072-3.
PEARSON, E. S. & HARTLEY, O. H. (Eds.) (1972). Biometrika tables for statisticians Vol. 2. Cambridge: University Press.
PEART, J. N. (1968). Lactation studies with Blackface ewes and their lambs. *J. Agric. Sci.* **70,** 87-94.
PETRUSEVICS, P. M. & DAVIDSON, T. W. (1975). Biotelemetry system for cattle behaviour studies. *Proc. Inst. Radio Electron. Eng. Aust.* **36,** 72-6.
POND, F. W. & PEARSON, H. A. (1971). Freeze branding cattle for individual identification. *J. Range Mgmt.* **24,** 466-7.
PRESTON, T. R. & WILLIS, M. B. (1970). Intensive beef production. Oxford: Pergamon Press.
RADFORD, H. M., WATSON, R. H. & WOOD, G. F. (1960). A crayon and associated harness for the detection of mating under field conditions. *Aust. Vet. J.* **36,** 57-66.
RAYMOND, W. F. (1948). Evaluation of herbage for grazing. *Nature, UK* **161,** 937-8.
RAYMOND, W. F. (1959). The nutritive value of herbage. *In* The measurement of grassland productivity. Ed. J. D. Ivins, London: Butterworths, pp. 156-64.
RAYMOND, W. F. (1969). The nutritive value of forage crops. *Adv. Agron.* **21,** 1-108.
RAYMOND, W. F., HARRIS, C. E. & HARKER, V. G. (1953). Studies on the digestibility of herbage. II. Effect of freezing and cold storage of herbage on its digestibility by sheep. *J. Br. Grassl. Soc.* **8,** 315-20.
RAYMOND, W. F., KEMP, C. D., KEMP, A. W. & HARRIS, C. E. (1954). Studies in the digestibility of herbage. IV. The use of faecal collection and chemical analysis in pasture studies (b) Faecal index methods. *J. Br. Grassl. Soc.* **9,** 69-79.
RAYMOND, W. F. & MINSON, D. J. (1955). The use of chromic oxide for estimating the faecal production of grazing animals. *J. Br. Grassl. Soc.* **10,** 282-96.
RAYMOND, W. F., MINSON, D. J. & HARRIS, C. E. (1956). The effect of management on herbage consumption and selective grazing. *Proc. VIIth Int. Grassl. Congr.* pp. 123-33.
REARDON, T. F. (1969). Relative precision of the tritiated water and slaughter techniques for estimating energy retention in grazing sheep. *Anim. Prod.* **11,** 453-60.
REID, J. T., WOOLFOLK, P. G., RICHARDS, C. R., KAUFMANN, R. W., LOOSLI, J. K., TURK, K. L., MILLER, J. I. & BLASER, R. E. (1950). A new indicator method for the determination of digestibility and consumption of forages by ruminants. *J. Dairy Sci.* **33,** 60-71.
REID, J. T., WOOLFOLK, P. G., HARDISON, W. A., MARTIN, C. M., BRUNDAGE, A. L. & KAUFMANN, R. W. (1952). A procedure for measuring the digestibility of pasture forage under grazing conditions. *J. Nutr.* **46,** 255-69.
REID, R. L. & MILLS, S. C. (1962). Studies on the carbohydrate metabolism of sheep. XIV. The adrenal response to psychological stress. *Aust. J. Agric. Res.* **13,** 282-95.

ROBINSON, J. J., FOSTER, W. H. & FORBES, T. J. (1968). An assessment of the variation in milk yield of ewes determined by the lamb-suckling technique. *J. Agric. Sci.* **70,** 187-94.

ROBINSON, J. J., FOSTER, W. H. & FORBES, T. J. (1969). The estimation of the milk yield of a ewe from body weight data on the suckling lamb. *J. Agric. Sci.* **72,** 103-7.

ROYAL, W. M. (1968). Equipment for collection of faeces from sheep. *Proc. Aust. Soc. Anim. Prod.* **7,** 450-4.

RUOHOMÄKI, H. (1975). Estimation of carcass characteristics in young beef cattle. *J. Sci. Agric. Soc. Finland.* **47,** 385-444.

RUSSEL, A. J. F., DONEY, J. M. & GUNN, R. G. (1969). Subjective assessment of body fat in live sheep. *J. Agric. Sci.* **72,** 451-4.

RUTTER, N. (1968). Time lapse photographic studies of livestock behaviour outdoors on the College Farm, Aberystwyth. *J. Agric. Sci.* **71,** 257-65.

RYLEY, J. W. & GARTNER, R. J. W. (1962). Drought feeding studies with cattle. 7. The use of sorghum grain as a drought fodder for cattle in late pregnancy and early lactation. *Queensl. J. Agric. Sci.* **19,** 309-30.

SCHNEIDER, B. H., SONI, B. K. & HAM, W. E. (1955). Methods for determining consumption and digestibility of pasture forages by sheep. *Wash. Agric. Exp. Sta. Tech. bull. No.* 16.

SEARLE, T. W. (1970). Body composition in lambs and young sheep and its prediction *in vivo* from tritiated water space and body weight. *J. Agric. Sci.* **74,** 357-62.

SEARS, P. D. & GOODALL, V. C. (1942). Apparatus for the collection of urine and faeces from grazing sheep. *N.Z. J. Sci. Tech.* **23**(A), 301-4.

SLATER, J. & JONES, R. J. (1971). Estimation of the diets selected by grazing animals from microscopic analysis of the faeces—a warning. *J. Aust. Inst. Agric. Sci.* **37,** 238-40.

SOUTHCOTT, W. H., ROE, R. & NEWTON TURNER, H. (1962). Grazing management of native pastures in the New England region of New South Wales. II. The effect of size of flock on pasture and sheep production with special reference to internal parasites and grazing behaviour. *Aust. J. Agric. Res.* **13,** 880-93.

SPARKS, D. R. & MALECHEK, J. C. (1968). Estimating percentage dry weight in diets using a microscopic technique. *J. Range Mgmt.* **21,** 264-5.

SQUIRES, V. R., DAWS, G. T. & BAWDEN, R. A. (1969). An automatic recording unit for photographic studies of behaviour of domestic animals. *J. Biol. Photogr. Assoc.* **37,** 188-98.

STEVENSON, A. E. (1962). Measurement of feed intake by grazing cattle and sheep. VIII. Some observations on the accuracy of the chromic oxide technique for the estimation of faeces output of dairy cattle. *N.Z. J. Agric. Res.* **5,** 339-45.

STEVENSON, A. E. & CLARE, N. T. (1963). Measurement of feed intake by grazing cattle and sheep. IX. Determination of chromic oxide in faeces using an Auto Analyzer. *N.Z. J. Agric. Res.* **6,** 121-6.

STEWART, D. R. M. (1967). Analysis of plant epidermis in faeces: a technique for studying the food preferences of grazing herbivores. *J. Appl. Ecol.* **4,** 83-111.

STEWART, D. R. M. & STEWART, J. (1970). Feed preference data by faecal analysis for African plains ungulates. *Zool. Afr.* **15,** 115-219.

STOBBS, T. H. (1970). Automatic measurement of grazing time by dairy cows on tropical grass and legume pastures. *Trop. Grassl.* **4,** 237-44.

STOBBS, T. H. (1973a). The effect of plant structure on the intake of tropical pastures. I. Variation in the bite size of grazing cattle. *Aust. J. Agric. Res.* **24,** 809-19.

STOBBS, T. H. (1973b). The effect of plant structure on the intake of tropical pastures. II. Differences in sward structure, nutritive value, and bite size of animals grazing *Setaria anceps* and *Chloris gayana* at various stages of growth. *Aust. J. Agric. Res.* **24,** 821-9.

STOBBS, T. H. & BRETT, D. J. (1972). The effect upon the fatty acid composition of milk of feeding tropical pastures to Jersey cows. *Proc. Aust. Soc. Anim. Prod.* **9,** 297-302.

STOBBS, T. H. & COWPER, L. J. (1972). Automatic measurement of the jaw movements of dairy cows during grazing and rumination. *Trop. Grassl.* **6,** 107-12.

STORR, G. M. (1961). Microscopic analysis of faeces, a technique for ascertaining the diet of herbivorous mammals. *Aust. J. Biol. Sci.* **14,** 157-64.

STORRY, J. E. (1970). Reviews of the progress of dairy science: Ruminant metabolism in relation to the synthesis and secretion of milk fat. *J. Dairy Res.* **37,** 139-64.

STOUFFER, J. R. (1969). Techniques for the estimation of the composition of meat animals. *In* Techniques and procedures in animal science research. New York, Albany: American Society of Animal Science, pp. 207-19.

STREET, J. C., BUTCHER, J. E. & HARRIS, L. E. (1964). Estimating urine energy from urine nitrogen. *J. Anim. Sci.* **23,** 1039-41.

STREETER, C. L. (1969). A review of techniques used to estimate the *in vivo* digestibility of grazed forage. *J. Anim. Sci.* **29,** 757-68.

TAN, T. N., WESTON, R. H. & HOGAN, J. P. (1971). Use of ^{103}Ru-labelled tris (1, 10-phenanthroline) ruthenium (II) chloride as a marker in digestion studies with sheep. *Int. J. Appl. Radiat. Isot.* **22,** 301-8.

TAYLER, J. C. (1954). Technique of weighing the grazing animal. *Proc. Br. Soc. Anim. Prod.* **1954,** pp. 3-16.

TILLEY, J. M. A. & TERRY, R. A. (1963). A two-stage technique for the *in vitro* digestion of forage crops. *J. Br. Grassl. Soc.* **18,** 104-11.

TORELL, D. T. (1954). An esophageal fistula for animal nutrition studies. *J. Anim. Sci.* **13,** 878-84.

TORELL, D. T., BREDON, R. M. & MARSHALL, B. (1967). Variation of esophageal fistula samples between animals and days on tropical grasslands. *J. Range Mgmt.* **20,** 314-6.

TREACHER, T. T. (1970). Apparatus and milking techniques used in lactation studies with sheep. *J. Dairy Res.* **37,** 289-95.

TYRELL, H. F. & REID, J. T. (1965). Prediction of the energy value of cow's milk. *J. Dairy Sci.* **48,** 1215-23.

ULYATT, M. J. (1972). Influence of experimental design in measuring the voluntary intake of grazing sheep. *Proc. N.Z. Soc. Anim. Prod.* **32,** 85-6.

UNDERWOOD, E. J. (1966). The mineral nutrition of livestock. Misc. Joint Publ. FAO/CAB. Aberdeen: Central Press.

VAN DER HONING, Y., STEG, A., & VAN ES., A. J. H. (1977). Feed evaluation for dairy cows: tests on the system proposed in the Netherlands. *Livestock Prod. Sci.* **4,** 57-67.

VAN DYNE, G. M. & HEADY, H. F. (1965). Botanical composition of sheep and cattle diets on a mature annual range. *Hilgardia* **36,** 465-92.

VAN DYNE, G. M. & TORELL, D. T. (1964). Development and use of the esophageal fistula: A review. *J. Range Mgmt* **17,** 7-19.

VAN ES, A. J. H. & VOGT, J. E. (1959). Separate collection of faeces and urine of cows. *J. Anim. Sci.* **18,** 1220-3.

VAN NIEKERK, B. D. H., BENSADOUN, A., PALADINES, O. L. & REID, J. T. (1963). A study of some of the conditions affecting the rate of excretion and stability of creatinine in sheep urine. *J. Nutr.* **79,** 373-80.

VAN SOEST, P. J. (1967). Development of a comprehensive system of feed analyses and its application to forages. *J. Anim. Sci.* **26,** 119-28.

VAN SOEST, P. J. (1975). Physico-chemical aspects of fibre digestion. *In* Digestion and metabolism in the ruminant; Proc. IVth Int. Symp. Ruminant Physiol. Eds. I. W. McDonald & A. C. I. Warner, University of New England Publications Unit, pp. 351-65

VAN SOEST, P. J. & MCQUEEN, R. W. (1973). The chemistry and estimation of fibre. *Proc. Nutr. Soc.* **32,** 123-30.

VAN SOEST, P. J., WINE, R. H. & MOORE, L. A. (1966). Estimation of the true digestibility of forages by the *in vitro* digestion of cell walls. *Proc. Xth Int. Grassl. Congr.*, pp. 438-41.

VICKERY, P. J. & HEDGES, D. A. (1972a). Mathematical relationships and computer routines for a productivity model of improved pasture grazed by Merino sheep. *Aust. CSIRO Anim. Res. Lab. Tech. Pap. No.* 4.

VICKERY, P. J. & HEDGES, D. A. (1972b). A productivity model of improved pasture grazed by Merino sheep. *Proc. Aust. Soc. Anim. Prod.* **9,** 16-22.

WALKER, D. E. (1962). Suckling and grazing behaviour of beef heifers and calves. *N. Z. J. Agric. Res.* **5,** 331-8.

WALLACE, L. R. (1948). The growth of lambs before and after birth in relation to the level of nutrition. *J. Agric. Sci.* **38,** 243-302.

WALLACE, L. R. (1956). The intake and utilization of pasture by grazing dairy cattle. *Proc. VIIth Int. Grassl. Congr.*, pp. 134-45.

WATSON, S. J. & HORTON, E. A. (1936). Composition, digestibility and nutritive value of samples of grassland products. *J. Agric. Sci.* **26,** 142-54.

WHEELCOCK, J. V. & DODD, F. H. (1969). Non-nutritional factors affecting milk yield in dairy cattle. *J. Dairy Res.* **36,** 479-93.

WHITELAW, F. G. (1974). Measurement of energy expenditure in the grazing ruminant. *Proc. Nutr. Soc.* **33,** 163-72.

WILLIAMS, C. H., DAVID, D. J. & IISMAA, O. (1962). The determination of chromic oxide in faeces samples by atomic absorption spectrophotometry. *J. Agric. Sci.*, **59,** 381-5.

WILLIAMS, O. B. (1969). An improved technique for indentification of plant fragments in herbivore faeces. *J. Range Mgmt.* **22,** 51-2.

WILLIAMS, O. B. & CHAPMAN, R. E. (1966). Additional information on the dye-banding technique of wool growth measurement. *J. Aust. Inst. Agric. Sci.* **32,** 298-300.

WILLIAMS, W. T. (1975). Pattern analysis in field experiments. *In* Development in field experiment design and analysis. Eds. V. J. Bofinger & J. L. Wheeler. *Commonw. Bur. Pastures Field Crops, Hurley, Berkshire, Bull.* 50, pp. 107-17.

WILSON, A. D., WEIR, W. C. & TORELL, D. T. (1971). Comparison of methods of estimating the digestibility of range forage and browse. *J. Anim. Sci.* **32,** 1046-50.

WINKS, L. & LAING, A. R. (1972). Urea, phosphorus and molasses supplements for grazing beef weaners. *Proc. Aust. Soc. Anim. Prod.* **9,** 253-7.

WODZICZA-TOMASZEWSKA, M. & BINGHAM, M. L. (1968). The tattooed patch technique for measurement of wool growth. *N.Z. J. Agric. Res.* **11,** 943-7.

YATES, N. G., MACFARLANE, W. V. & ELLIS, R. (1971). The estimation of milk intake and growth of beef calves in the field by using tritiated water. *Aust. J. Agric. Res.* **22,** 291-306.

YOUNG, B. A. (1970). Application of the carbon dioxide entry rate technique to measurement of energy expenditure by grazing cattle. *Eur. Assoc. Anim. Prod. Publ. No.* 13, pp. 237-41.

YOUNG, B. A. & CORBETT, J. L. (1972a). Maintenance energy requirement of grazing sheep in relation to herbage availability. I. Calorimetric estimates. *Aust. J. Agric. Res.* **23,** 57-76.

YOUNG, B. A. & CORBETT, J. L. (1972b). Maintenance energy requirement of grazing sheep in relation to herbage availability. II. Observations on grazing intake. *Aust. J. Agric. Res.* **23,** 77-85.

YOUNG, B. A. & WEBSTER, M. E. D. (1963). A technique for the estimation of energy expenditure in sheep. *Aust. J. Agric. Res.* **14,** 867-73.

YOUNG, J. S. (1974). Reproduction and nutrition in the beef herd. *Proc. Aust. Soc. Anim. Prod.* **10,** 45-54.

Chapter 8

PATTERN SEEKING METHODS IN VEGETATION STUDIES

M. B. Dale

I. INTRODUCTION

Classical statistical methods of data analysis are concerned either with the estimation of some parameter value or with a test of some hypothesis. In contrast, pattern seeking methods are concerned first to organize observational data so as to generate hypotheses, and second to do this by ordering the hypotheses so that potentially interesting ones may be tested first. Pattern seeking methods thus form an aid to efficient utilisation of experimental resources.

In any set of observational data there are manifold patterns. We select between them on grounds of elegance, of parsimony and, most important, of utility. But utility requires some context for evaluation, a context provided by the user and not the data; we are actually concerned with patterns-for-agents (MacKay, 1969). Thus, each set of data should strictly demand its own unique method of analysis, but such profusion is not demanded of the methods to be discussed here. Humans find few patterns cogent for their use, and it is with some of these that we are concerned. However, extensive practical detail is not appropriate here, and the reader is referred to the works of Sneath and Sokal (1973), Anderberg (1973), Orlóci (1975), Clifford and Stephenson (1975) and Williams (1976) for more complete descriptions.

Methods of pattern seeking are often characterised as objective methods. In the limited sense that they are defined by provable algorithms, or more often by specific programs for a computer, they are indeed repeatable, and hence objective. However, a considerable amount of subjective decision making is required before the pattern-seeking methods can be used. We must decide which items are to be organised, how to describe them and which kind of organisation is currently of most interest to us. There is no guarantee that long and involved mathematical calculations produce results of automatic worth, and not all properties of models can be unquestioningly transferred to the real world. Objectivity may perhaps redirect subjective decisions to profitable paths, but should not remove it.

Agronomic and ecological studies are concerned with interactions between organisms and their environment. For practical purposes we concentrate initially on one of these, seeking to identify patterns in it which can be interpreted by using information from the other. Thus, following the arguments of Lambert and Dale (1964), we might first examine plants, seeking to relate the patterns found in them to environmental variation later. Alternatively, we can reverse the roles of plant variation and environmental variation, and adopt the view of Whittaker (1973). It would be most appropriate if we regarded the whole problem as a dialogue between organism and environment to which each contributes, yet to describe which each, alone, is insufficient. Methods for examining such interaction are undeveloped, and it remains profitable, and interesting, to examine one part of the dialogue in detail first. In this chapter we shall adopt the view that the organisms form a more convenient starting point.

All of the methods we shall examine start with data in the form of a set of items, for each of which we have a list of property values. In some of the methods this form of the data is implied rather than being explicitly required. Since we need such a set of data, the first task is to define the items and to identify the properties.

II. DATA COLLECTION

1. Item definition

In some studies the items which must be organised would seem self-evident. Individual organisms, enterprises, manuscripts and languages clearly form items of interest and can be easily identified. In vegetation or ecosystem studies there is no obvious candidate for the basic item. It could range from an individual leaf in a study of litter decomposition, to a subcontinental area in a study of biogeography. The traditional method for agronomic and vegetation studies has been to select quadrat samples spaced over the area of interest and use descriptions of these for pattern seeking. With such a system it is necessary to determine if the quadrats should be selected on locational or vegetational criteria, and to determine their size, shape and the pattern of their distribution over the area of interest. The records may cover a single period of time or several periods. Various treatments might be applied, and in some cases physiological states encountered may further distinguish items. For example, we may wish to regard a single animal as several different items, depending on whether it is juvenile, maiden, pregnant, lactating or barren.

The pattern-seeking methods themselves do not formally constrain what may be regarded as an item. This identification is limited only by the ingenuity of the user and his eloquence in persuading his peers that the selected items are appropriate in context. There are, of course, some constraints imposed by the methods and still more introduced by lack of adequate development of appropriate computer programs. As examples, we can take the item definition of Williams *et al.* (1969a) who defined a sample item as a specified tree and its k nearest neighbours. The size of the sample thus depends on k and on the tree density for fixed k, while the shape was implicitly circular. The distribution was exactly that of the trees themselves, since each tree in turn was used as the identifying individual. It would be possible to modify sample shape as well in accordance with some vegetation property if this seemed necessary. Stephenson, Williams and Lance (1970) found that local variation was so high in marine sampling that local averaging, or smoothing, was desirable.

Whatever the items, though, they must remain meaningfully comparable. In one study several small plots were described in terms of native vegetation. The vegetation and topsoil were then removed, the underlying parent rock subjected to extractive mining and the topsoil replaced. Moreover, the replacement did not conserve either position or topography, so that no study of succession during revegetation could reasonably use the pre-mining information.

2. Property definition

When determining the items, the major problem is concerned with representativeness, i.e. with obtaining information about unknown patterns without excessive bias, or contamination from uninteresting variation. In turning to the description of the items, the major problem becomes one of relevance; in order to interpret some pattern we need to be able to relate it to other information, and thus assess its utility. In vegetation studies, we may measure a single property without regard for heterogeneity within the unit item. For example, much of the work on productivity has emphasised the importance of biomass, without further qualification. In most cases, though, we need to account for internal, spatial and temporal variation in the vegetation of our item samples. The traditional means of doing this is to identify taxonomic categories within the vegetation. Agronomists have often split or lumped taxa; instead of using the species category, they might use cultivars for ' interesting ' parts, while lumping other parts into categories such as ' weeds ' or ' other grasses '. Dale and Clifford (1976) have initiated an examination of the results of such practices, although it is clear that lumping may involve irretrievable loss of information, while splitting may demand an inordinate amount of recording time.

Most ecological studies have been less adventurous and have stayed with the species as a basic unit for partitioning plant material. There are, of course, alternatives which do not use taxonomic categories. Webb *et al.* (1970, 1971) have used structural and life-form categories to good effect, and similar concepts have been employed by Lambert (1972) and Dansereau, Buell and Dagon (1966). The extreme case of such an approach is the work of Grabau and Rushing (1968), who described in considerable detail the three-dimensional distribution of plant material. This detail was specifically needed for their problems, which concerned difficulty of penetration of vegetation by sight, soldier, projectile or vehicle. This illustrates well the need to select the properties to be recorded by their relevance of the problem being considered.

After choosing the partition of vegetation, we must decide how it is to be measured, and in what scale of measure the recording is to be made. With species, a range of measures can be employed which roughly forms a series of increasingly detailed measures (see also Chapter 3). The series runs :

presence : a record of the existence of some property. Often called *qualitative*;
density : number per unit area;
valence or frequency : a measure incorporating both density and local distribution pattern;
cover : frequency compounded with size and shape in vertical projection;
cover repetition : separation of vertical segments of cover and approximating volume;
biomass : adding the variation in material density to volume.

There has been much discussion concerning which of these measures should be used, although it should be immediately obvious that there is no unique measure suitable for all problems. Smartt, Meacock and Lambert (1974) have examined the comparability of several such measures. Unfortunately, their results are not incontestable, owing to the possibility of non-linear relationships between the measures. Furthermore, the usefulness of a particular measure is related to the precision of estimate and to amount and rapidity of fluctuation in the value. As a compromise, several scales of cover-abundance which give some coarse measure of amount of plant material, have been proposed. Such scales can be used in estimations which involve little more time than a simple presence list.

When categories other than species are used, it may be possible to use the same kind of measures, but often various properties must be recorded in a variety of scales of measure. Such problems are common in numerical taxonomy and Lance (1970) has listed many of the possibilities. The important types are :

Disordered multistate: the property has a series of disjoint nominal values, for example, ' married state ' has the values ' unmarried ', ' married ', ' widowed ', ' divorced '. Usually, the values are exclusive but occasionally items may possess two or more values giving non-exclusive categories. If there are only two states the property is often called a binary character.

Ordered multistates : Here the values still represent classes, but they may be ordered in magnitude. Age as a property might have the values ' infant ', ' juvenile ', ' adult ', and ' senile ', which can clearly be ordered.

Numeric : Here we finally obtain the precision of real numbers, which can be subjected to arithmetic operations. Examples would include biomass, height and density.

Ideally, of course, all the various data types would be uniformally treated in their several categories by the pattern seeking method. Unfortunately, this is not possible and no computational method exists which will handle all types. This means that the user of any pattern seeking method is often forced to consider recoding some values in order to avoid computational problems. At the same time, he may employ various transformations of his data to improve its relevance. Perhaps the commonest of such transformations has been to zero mean and unit variance (c.f. Noy-Meir, 1973a; Greig-Smith, Austin & Whitmore 1967; Lacoste & Roux, 1971, 1972) and various attempts at reducing correlation between various properties. The reasons for employing such transformations have not always been clear, although the use of a logarithmic transform to mitigate the effects of outlying values has often proved valuable. Even worse is the inclusion of a transformation in a computer program, for the user may not be aware of its occurrence, and unable to redress its effects. It may be desirable, but it may equally reflect the local biases of the program author.

Two further questions require some attention. In real data, missing values are not an uncommon occurrence, whether they are unrecorded, inapplicable or just lost. A good method will attempt to redress such losses, although their introduction should be eschewed. Finally, we might avoid recording properties of items and instead record directly interrelationships between items. Social science has made much use of paired comparisons where such direct estimation is made by a judge. Ecologically, we might record species growing ' over ', ' under ', ' next to ', ' touching ', ' on ' (epiphytic) or ' after '. Such relationships might be calculable, given detailed three-dimensional information on the individual plants, but they can equally be measured directly. Williams *et al.* (1969b) have used the ' after ' relationship in following successional sequences. This does not exhaust the possibilities, of course. One intriguing example occurred in classification of grammars. Carvell and Svartvik (1969) recorded whether or not a particular transformation was applicable to a given sentence. Thus, trivially, we might ask if an area could be ploughed for some fixed cost and the resulting answer used as a partial description of the area. As with item selection, the properties used to describe an item can be adapted to serve many purposes and careful selection may considerably enhance the effectiveness of pattern seeking.

III. DATA ANALYSIS

1. Strategies of analysis

There are two main streams apparent in pattern seeking methodology, although recent work (Dale & Anderson, 1973; Dale & Webb, 1975) has somewhat blurred the distinction. It is, however, still convenient to distinguish *ordination* methods from *classification* methods. In ordination, all the data are represented in some simplified manner, and patterns are sought in this simplified representation. We may then reintroduce the specific complexities of the problem to a simplified solution. There are formal methods for recognizing analogies or generalising, which have no obvious geometric interpretation, but most pattern seeking ordination methods either represent the data, or a function of it, in a low-dimensional Euclidean space, or seek a simple ordering of items, which technically involves rank-order or isotonic space.

The alternative approach is to subdivide the data into sub-groups, such that members of the same sub-group are ' alike ' and members of different sub-groups are ' unlike '. Having identified sub-groups we first seek to explain them and then recombine them into a hopefully convincing pattern for all the data.

Both ordination and classification methods use models. In ordination we seek a low-dimensional representation and use a measure of adequacy of fit to determine if the representation is sufficient. In classification we need a measure of ' likeness ' which can be regarded again as a measure of adequacy of fit of some model to each sub-group. In ordination the response to an inadequate fit is to complicate the model, in classification it is to subdivide the items.

A perusal of the textbook of Sneath and Sokal (1973) will not reveal much discussion of the models used. Instead, there is a considerable account of a large number of similarity measures. These are measures between pairs of items which are functions of the property values for the items and seek numerically to represent the degree of ' likeness ' of any pair of items. Most widely used pattern seeking methods are based on similarity calculations, and before considering what similarity measures represent, we must give a cursory examination of the techniques used to calculate them.

The first problem is whether all properties should be included in the calculation, or if some should, permanently or temporarily, be ignored. Webb *et al.* (1967) showed that the basic patterns of their vegetation could be recovered from only a small proportion of the species used in the description—in fact only 12 percent was needed. They recommended that very rare species could be ignored. In contrast, Goodall (1969) has strongly emphasised the role of rarities and his probabilistic similarity measure strongly weights uncommonness. The ecological merits of these opposing viewpoints are still being debated. There is, however, a plausible case for not employing all available information in a single classification, especially where the properties themselves fall into disjoint sets. Separate analyses of each set may be much more profitable under these circumstances, since the results may be compared and contrasted.

Ignoring properties entirely is simply an extreme example of weighting. The user can, if he desires, differentially weight various properties so that their contribution to a similarity measure is enhanced or attenuated. In general, the problem here is to determine the source of the weighting. It can be based on the data itself (Williams, Dale & MacNaughton-Smith, 1964), although this is computationally demanding. In some cases the existence of one property is dependant on the value of another property; it is hard to find the hairs on the leaves of a leafless plant! Williams (1969) has shown how much dependence can be mitigated by appropriate weighting. Finally, the source of the weightings may be the experience of the user. While such belief statements are a necessary component of selection of properties, they are usually regarded as insufficiently precise to provide adequate grounds for differential weighting. The difficulty, of course, is the possibility of circularity, where the subjective weighting produces patterns which confirm the weighting. On the other hand, the user may well include considerable amounts of other information in the establishment of his subjective weights and these beliefs may, and perhaps should, be incorporated in the analysis. Weighting is a very vexed question, and the author can only indicate that in general he does not find differential weighting overly successful.

Having chosen a set of properties and their weights, the selection of an appropriate similarity measure is necessary. This is still a craft skill, with different authors recommending different coefficients. In my own work, the following rules of thumb are used, although some users do select other alternatives. If the data are primarily multistate, with few or no numeric values, then information measures are suitable and effective and the method of Williams, Lambert and Lance (1966) and its modifications are appropriate. Where the data are primarily numeric and no property is distinguished by extreme outliers, i.e., the frequency distribution for any property is roughly symmetric, then the method of Burr (1968) has proved valuable. Where extreme outliers do occur, the Canberra metric of Lance and Williams (1967) is probably preferable. Finally, for ecological data in frequency or percentage form, the CENTPERC method of Dale (1971) and especially the extended form due to Williams (1972, 1973), has proved effective. These are, of course, personal empirical rules but they have been developed from analyses of real data.

2. Models and similarities

Having obtained our similarities, we can now return to the question of the relationship between the similarity measure and the model whose adequacy of fit is in question. For binary data, most similarity measures resolve into a counting of mismatches and it is difficult to see what else could be done in this case. Indeed, most similarity measures are of an analogous form, which Wishart (1969) called the minimal variance form. The model implied by such coefficients is simply one where, within any group of items identified in the analysis, the variance for each property will be minimal and ideally will be zero. Note that this makes the correlation between properties within groups indeterminate. It is not necessary to accept the minimal variance model, but almost all pairwise similarity measures reflect this model. Any pattern analysis of such similarities is then equally concerned with minimal variance. Yet it is not at all clear that such a model is desirable for all pattern seeking and there is evidence, discussed in Dale and Anderson (1973), suggesting that other models are ecologically more appropriate.

The possibility that pairwise similarity was not fundamental first arose in ordination studies. For, while similarity can be sensibly measured between ' like ' items, very disparate items can all be regarded as equally ' unalike ' one another. In effect, if the difference is too great it simply becomes no similarity at all, irrespective of the reasons why. Thus, ordination methods based on local neighbourhood criteria have long been sought. In classification the questioning began because it is not clear that the similarities calculated initially should remain inviolate throughout an analysis, and because in the calculation of similarities involving zero values, as in much ecological work, it is not clear how far two zero values are really the same ecologically. Both these questions reflect an interest in the context of the measurement of similarity and how that context might be changed during analysis. Jardine and Sibson (1971) have shown that accepting the initial similarities can have attractive mathematical properties, although it is not clear that this also applies to the results given to the user. However, rejection of inviolateness necessitates the specification of rules determining the changes to be made. Williams, Clifford and Lance (1972) have investigated how some coefficients react to group size during an analysis.

Finally, similarity measures have not provided adequate indication of interesting groups; that is, they do not provide a stopping rule. Most methods of classification involve joining or splitting sub-groups. Yet, while the decision as to which groups should join or split can be made on the basis of similarity between groups, the decision to join, to split or to accept the present groups is not a simple function of the similarity.

These difficulties disappear if the idea of adequate model fit is accepted as primary with similarity as a derived characteristic. For, if the model fits, nothing need be done, whereas if it does not fit, some operation must be performed to search for a better solution. Similarity can be regarded simply as the change in the same model when two items or groups are combined, and is thus an (i, j, k) measure in the sense of Williams, Lambert and Lance (1966). However, in some cases it is easier to calculate the change in fit without explicitly fitting the model at all; indeed, in some cases the model implied has never been fully examined. Thus, similarity calculation is a computational convenience.

It should be noted that the concept of model fitting applies to any number of items. One of the most awkward features of similarity measures was that, if they were defined for pairs of items, they had to be arbitrarily defined when groups of items became involved. As we shall see in the next section, such group to group similarities are found in all classification methods.

3. Search strategies and classification

There are various ways of categorising the methods used in searching for classes. Some methods seek groups of items to define a class yet any item may form part of several groups (Jardine & Sibson, 1968; Yarranton *et al.,* 1973), but most methods seek exclusive groups of items. Within the exclusive grouping methods we may either search directly for some fixed number of classes or clusters, or obtain these by forming a hierarchy of classes, with nested groups of subclasses within classes. Hierarchical methods may be divisive, working primarily by sub-dividing groups, or agglomerative, working primarily by joining sub-groups into larger groups. Finally, divisive methods may be polythetic, using all properties, or monothetic, using one property, specially selected to characterise the division being made, rather like a dichotomous key. Hierarchical methods actually relate more closely to Bunge's (1969) level structures than to hierarchies proper, but regrettably this is not reflected in common usage. Williams and Dale (1965) gave more detailed definitions of these categories. All these alternatives aim at one objective, to improve the overall fit of some model to the subclasses of the data. Overlapping clusters, or clumps, have not proved popular and may be disregarded here.

Non-hierarchic methods can either stand alone, or be linked to some hierarchic method which serves to generate some initial groupings. Most of these methods start from some initial partition of the items in k groups. Individual items are then moved between groups, in order to increase the fit of the model to all sub-groups. The techniques differ principally in the manner of initialisation and the rules governing the transfer process. Some methods allow groups to be split further, or lumped together if this is necessary. Non-hierarchic methods have one special property when used with the minimal variance model, for in that case an optimal solution can be found. Vinod (1969) and Koontz, Narendra and Fukunaga (1975) have indicated how this can be accomplished. However, both these methods still require the specification of the number of groups to be found. It is still not clear that exploration over various numbers of groups is possible.

Much more attention ecologically has been given to hierarchic methods, and in particular, to polythetic agglomerative techniques. Only recently has attention been focussed again on divisive methods, and many users will be surprised perhaps to find that polythetic agglomerative methods are not the entirety of classification methodology.

Polythetic agglomerative methods start by calculating similarities between all pairs of items. This calculation is the weakest link, for it becomes unfeasible for more than a few hundred items, using present techniques. Having obtained all the pairwise similarities, all the methods proceed in the following manner :

(i) Find the most similar pair of items or groups;

(ii) Fuse this pair, and recalculate similarities of other items to this composite;

(iii) If everything is not in one group go back to (i). Otherwise stop.

This, of course, does not provide the detail. There may be no unique maximal similarity, so the pair for the next fusion may not be unambiguously selected. Jardine and Sibson (1971) have made much of this difficulty. In most cases allowing multiple fusions (Orloci, 1966) or reference to other similarities (Williams *et al.*, 1971b) resolves the problem and the residuum of unresolved cases usually involves several completely identical items which will all fuse anyway. If all fails, then the groups will depend on the order of entry of items into the search for maximal similarity.

The other problem is how to calculate similarities from groups to other groups or other items. For methods explicitly fitting a model there is no problem; however, when similarity measures only are available, Lance and Williams (1966) have shown that most proposals form part of a general expression appropriately parametrised. Williams, Clifford and Lance (1972) have explored the effects of various rules and shown that these may well alter similarities during analysis. This can be useful to clarify demarcation of groups, yet Jardine and Sibson (1971) noted some undesirable properties. This argument continues.

Before considering divisive methods, there are a few unusual methods deserving note. Zahn (1971) used the similarities in a calculation to find a minimal spanning tree and then looked for links in the tree which could be broken because they were too large (cf. Jancey, 1974). Wishart (1969) felt that reliance on nearest neighbours was not warranted and used similarity to k^{th} neighbour to order his items. He then devised a grouping method which is one of the few which actually identifies automatically the number of groups needed. Diday (1974) has treated the problem of combining results of alternative non-hierarchic clusterings by employing an agglomerative technique.

4. Divisive search methods

Agglomerative methods spend much effort forming trivial groups at the start of the analysis, with major distinctions deferred until near the end of processing. Divisive methods start by finding major distinctions by successive dichotomous divisions. However, it is not possible to examine even all divisions into two classes for a large number of items, notwithstanding the heroic suggestion of Edwards and Cavalli-Sforza (1965). Instead, only a restricted subset can be tried, so that differences in divisive methods mostly concern selection of the set of putative divisions.

Polythetic methods have been devised by McQuitty (1968), MacNaughton-Smith *et al.* (1964), Noy-Meir (1973b) and by Lambert *et al.* (1973). These last have examined various methods in comparison with the optimal division for small sets of data, but the two methods selected as best approximations are both fairly laborious. Many of the methods are based on principal component analysis, and with increasing number of properties this will fail finally because of excessive computational requirements. Various sequential methods such as Uttley's (1970) INFORMON have seen little application. McQuitty's (1968) method relies on correlating the correlations of the correlations of the correlations etc., and will again fail if the number of properties is too large, as will techniques of Eigen, Fromm and Northouse (1974). However, in analysing data such as those received from ERTS sources these methods can be effective. Incidentally, it should be noted that agglomerative methods do not really perform any search at all, which may account for their poor performance in the tests of Lambert *et al.* (1973). Their results are dominated by the first few fusions made when the available information is minimal.

Monothetic divisive methods have been proposed by Goodall (1953), Williams and Lambert (1959), Lance and Williams (1968) and Crawford and Wishart (1967, 1968) with Hill, Bunce and Shaw (1975) providing a hybrid analysis using five indicators for each division. All these methods employ each property in turn to provide a subdivision, and select from this set the best available one. The two or more sub-groups defined are then independently divided further. In Goodall's and Williams and Lambert's methods, the appropriate property was selected directly, Goodall using the most frequent species, Williams and Lambert the species most highly correlated with all others.The Lance-Williams and Crawford-Wishart methods both involve model fitting, the former using an information model, the latter their ‘ group element potential ’.

All these methods conform closely to those used in heuristic search (Slagle, 1970; Pohl, 1969, 1970; Doran, 1969) and in branch and bound studies (Hall, 1971). Heuristic search has been widely used in robotics, theorem proving and game-playing programs, and efforts have been made to provide more effective means of searching. This might involve following several alternatives, by investigating the consequences of various alternatives at later stages (depth increase), or adding more alternatives to the properties (breadth increase). Pohl (1969) and Harris (1974) have examined the effects of errors on such searches and how they can be mitigated. Much of this work has made little impact on pattern-seeking methodology as yet.

The process of division into subgroups stops if the model fits all subgroups adequately. Thus, the termination is a function of model fit and this has been used to identify interesting or significant groups. Recently, Dale and Anderson (1973) and Dale and Webb (1975) have suggested other indications of interesting groups which do not rely solely on the measure of fit.

The assumption made in monothetic methods is that the groups sought will possess or lack some range of property values consistently and many users regard this as an untenable assumption. We can, however, regard the monothetic method as a quick and dirty approximation and try and clean up the results. In effect, we regard the monothetically derived groups as the initial partition for a non-hierarchical clustering. The problems with reallocation of this kind are when to apply and what exactly to do. We might reallocate after every division either between the two sibling groups just formed, or over all groups as Lambert *et al.* (1973) do in their MONIT program, we might delay reallocation until all division has been terminated, or we might use some indicator to suggest when and over what groups reallocation is required. Weir (1970) has constructed a program which reallocates over terminal groups. He allows groups to vanish if all their members are better placed elsewhere, but no group can further subdivide, whereas Boulton and Wallace (1968) provide for both vanishing and splitting. Weir also keeps track of pattern of movements to prevent items oscillating indefinitely and modifies the measure of fit so that it is less sensitive to group size than in the subdivision phase. For binary data, this overall approach provides a cheap and effective method of pattern finding.

5. Inverse and two-way analyses

Since monothetic methods are relatively cheap to use, if properly programmed, we can easily explore alternative strategies with such methods. Throughout, we have emphasised grouping of items, but by simply transposing the data we may equally well group properties in an Inverse Analysis. In practice, such grouping has proved ineffective, since it has been dominated by the commonness or rarity of species. This is primarily due to the use of the minimal variance model, which is ineffective for binary data and inappropriate for many other data types, in inverse analyses.

Lambert and Williams (1962) suggested examining the coincidence of item and property groups, and they suggested a method for identifying coincident groups. Dale (1964) has modified their proposal to use more directly the density of entries. It then becomes apparent that direct search for such two-way defined groups is possible and Tharu and Williams (1966) and Hartigan (1972) have both defined such methods, although neither is particularly attractive for other than binary data.

This does not exhaust the possibilities, for only the most tentative explorations have been made of models or search methods. There is, of course, a relationship between the model sought and the technique used to find it, but at present there is so little experience with any model other than minimal variance, that the required techniques are unknown.

6. Ordination, simplification, seriation (OSS)

Once similarities between items have been calculated, several analyses may be carried out relatively easily. Most ordination methods are dependent on the use of similarity measures, so that the adoption of an agglomerative approach to classification can be combined with an alternative ordination. In ordination we seek a simple representation of data, and progressively increase the complexity of the model until adequate fit has been obtained.

The first attempts to use ordination in pattern seeking were due to Hotelling (1933) who developed principal component analysis, and to various psychologists who developed the techniques of factor analysis (Harman, 1967). If we have n items described by m properties, these techniques seek a least squares approximation using p new properties. These new properties are defined as linear combinations of the original m properties, each of the new properties being orthogonal, that is linearly uncorrelated with any of the other new properties. Further, each of the p properties of components is associated with a measure of importance, so that they can be ordered, and trivial components eliminated. In general, p is usually not much less than m and the rejection of unimportant components is necessary to obtain much simplification. There is, of course, a risk that some of the apparently trivial components may contain real and interesting information, which may then be rejected, for the measure of importance is not necessarily a measure of utility or interest.

Principal component analysis uses the interproperty dispersion or correlation matrix, and with data of various types the definition of correlation is difficult. To avoid this, Gower (1966) proposed his principal co-ordinate analysis which is based on inter-item dissimilarity measures. Dissimilarity is simply the complement of similarity, and is often termed distance. Gower showed that such dissimilarities can be mapped to a closely approximating Euclidean representation, and in certain circumstances the results are identical with those obtained by component analysis. However, co-ordinate analysis is much more widely applicable.

Ecological examples of ordination can be found in Goodall (1954), Ivimey-Cook and Proctor (1967), Orlóci (1966), Greig-Smith, Austin and Whitmore (1967), Noy-Meir (1971, 1974) and Anderson (1971) among others. Beals (1973) has indicated that in many cases the models are simplistic ecologically and he questions the validity of their use. He was primarily concerned with non-linearity and questioned whether the linear models used in ordination were appropriate in ecological studies. Dale (1975) has suggested that at least two different objectives were apparent in ordination studies, one concerned with efficient simplification or data reduction, the other with the arrangement of items, and perhaps properties, in sequences (cf. Lange, 1968). For data reduction, linear additive models can be useful as in Noy-Meir and Anderson's (1971) multiple pattern analysis. For sequencing, such models do not seem as useful, but we shall discuss this later.

There is a further consideration ignored by both Beals and Dale. For linear additive models, the principal component and co-ordinate solutions have useful optimality properties. The non-linear alternatives mostly use complex iterative calculations which may converge to local rather than global optima. In effect, we need to compare the value of an optimal solution to a possibly inappropriate model with a possibly suboptimal solution for a more realistic model.

The non-linear alternatives proposed are quite varied in their assumptions. Kruskal's (1964) method uses rank-order information, Zinnes and Griggs (1974) assume a particular unimodal symmetric distribution, while Shepard and Carroll (1966) use only local similarity and seek models with only topological similarity. All these methods, and there are others, involve somewhat arduous calculations.

One encouraging possibility lies in Hill's (1973) rediscovery of reciprocal averaging. This is primarily a sequencing or seriation method, since it seeks to order items and properties into a sequence, such that similar items are close in the sequence, as are similar properties. The advantage of the method is that it finds a unique globally optimal solution. Unfortunately, the concept of a sequence seems inherently one-dimensional and considerable practical and theoretical problems remain before an adequate multidimensional model can be employed.

7. Extrinsic and sequential patterns

Before concluding this brief account of pattern-seeking methods, two other classes must be considered. In extrinsic methods we are concerned with relationships between two, or more, sets of properties for a single set of items. In many cases multiple regression is appropriate but some other methods exist which can be useful. Williams and Gillard (1971) used ordination methods to simplify both sets of properties before employing canonical correlation analysis, an approach also adopted by Webb *et al.* (1971) and Williams *et al.* (1971a) with some success. The rejection of trivial information before cross relationships were investigated did seem, in these cases, to be a useful adjunct to the standard methods. Webster, Gunst and Mason (1974) and Mantel and Valand (1970) have used similar methods.

Extrinsic classificatory methods have been described by MacNaughton-Smith (1965), and of his suggestions, the multiple predictive analysis has been successfully used by Beeston and Dale (1975). The objective here is to maximise the relationship between 'treatment' variables and 'result' variables by subdivision based on a third, ' background ', set of variables. Beeston and Dale used original vegetation as background, clearance method as treatment and density of regrowth as result, but the definition of appropriate sets of variables is obviously problem-dependent. Since the analysis involves fitting a model for some sets of properties while using others to define classes, it still fits the search paradigm. By incorporating costs and values, an analogous classification method could search for cost/effective treatments to obtain some result. Similar approaches are apparent in the work of Furnival and Wilson (1974) in searching for best possible regressions, and of Sonquist, Baker and Morgan (1973), who considered various alternative models for specifying the treatment-result relationship.

Sequential patterns appear in two different guises. In one, techniques of classification and ordination are applied to problems where property values are recorded over some period of time, giving data in the form of items, properties and periods. The work of Williams and Edye (1974) on three-dimensional problems is apposite here. In the other, it is the recognition of recurrent patterns in space and time which is of interest. Work here includes that of Greig-Smith (1952), Kershaw (1960) and Noy-Meir and Anderson (1971), involving split-plot techniques to estimate scales of patterns and that of Morris (1972) and Notley (1970), who both defined self organising systems which can identify repetitive or recurring patterns in sequences of observations. This latter approach has been adapted to studies of soil profiles by Norris and Dale (1971). The methods of spectral analysis (see e.g. Jenkins & Watts, 1968) is also relevant here, but the results obtained are often incommensurate with the computation involved.

Both extrinsic and sequential methods are important, because they both examine interaction and the development of that interaction in time, and thus approach the goal of studying the environment/organism dialogue directly.

IV. CONCLUSIONS

At various places in the text, the reader may have noticed a certain discomfort with present techniques and methodology. The available methods can form useful tools in examining large amounts of data and exposing certain features which might otherwise be overlooked. However, the methods are certainly still in a process of development and the minimal variance model is perhaps the most obvious weakness. Indeed, it was dissatisfaction with this model which led Dale and Anderson (1973) to investigate models where a specific correlation structure provides the definition of an ideal group. Their two-parameter model links classification to ordination in a single technique, which seems to provide a more acceptable model both for ecological and taxonomic analyses. Furthermore, the symmetry of the model allows comparison of normal and inverse alternatives and yields various indicators which can aid in providing an efficient search.

Dissatisfaction with the models is not the entirety of the discontent, however. Pattern-seeking methods can be regarded as attempts to infer ' grammars ' for observed data; that is, they attempt to determine the rules governing the development of patterns. This syntactic preoccupation has masked the fact that the generation of hypotheses also involves the assignment of meaning to the patterns, i.e. the interpretation of the patterns in terms of other variables. This semantic component has been somewhat lost under the pile of algorithms and similarity measures and is only now beginning to receive attention. Yet, it is in the interpretation that our beliefs are most strongly apparent. There is as yet no universally accepted theory of inference, no means by which interpretation could be fully automated. Human users presently provide the only practicable inference makers, and until some feasible automated alternative is available, the only means of valuing the performance of pattern-seeking methods is to demand of the user a value judgement. It is to the provision of such *inferential* methods that pattern-making methodology must now turn.

V. REFERENCES

ANDERBERG, M. R. (1973). Cluster analysis for applications. New York: Academic Press.

ANDERSON, A. J. B. (1971). Ordination methods in ecology. *J. Ecol.* **59,** 713-26.

BEALS, E. W. (1973). Ordination: mathematical elegance and ecological naïveté. *J. Ecol.* **61,** 23-36.

BEESTON, G. R. & DALE, M. B. (1975). Multiple predictive analysis: a new management tool. *Proc. Ecol. Soc. Aust.* **9,** 172-81.

BOULTON, D. M. & WALLACE, C. S. (1968). A program for numerical classification. *Comput. J.* **13,** 63-9.

BUNGE, M. (1969). The metaphysics, epistemology and methodology of levels. *In* Heirarchical structures. Eds. L. L. Whyte, A. G. Wilson & D. Wilson. New York: American Elsevier Publishing Co., pp. 17-28.

BURR, E. J. (1968). Cluster sorting with mixed character types. I. Standardization of character values. *Aust. Comput. J.* **1,** 97-9.

CARVELL, H. T. & SVARTVIK, J. (1969). Computational experiments in grammatical classification. The Hague: Mouton.

CLIFFORD, H. T. & STEPHENSON, W. (1975). An introduction to numerical classification. New York: Academic Press.

CRAWFORD, R. M. M. & WISHART, D. (1967). A rapid multivariate method for the detection and classification of groups of ecologically related species. *J. Ecol.* **55,** 505-24.

CRAWFORD, R. M. M. & WISHART, D. (1968). A rapid classification and ordination method and its application to vegetation mapping. *J. Ecol.* **56,** 385-404.

DALE, M. B. (1964). The application of multivariate methods to heterogeneous data. Ph.D. Thesis, Univ. Southampton, UK.

DALE, M. B. (1971). Information analysis of quantitative data. *In* Statistical ecology. Eds. G. P. Patil, E. C. Pielou & W. E. Waters. Pennsylvania: State University Press, pp. 133-48.

DALE, M. B. (1975). On the objectives of methods of ordination. *Vegetatio* **30,** 15-32.

DALE, M. B. & ANDERSON, D. J. (1973). Inosculate analysis of vegetation data. *Aust. J. Bot.* **21,** 253-76.
DALE, M. B. & CLIFFORD, H. T. (1976). On the effectiveness of higher taxonomic ranks for vegetation analysis. *Aust. J. Ecol.* **1,** 37-62.
DALE, M. B. & WEBB, L. J. (1975). Numerical methods for the establishment of associations. *Vegetatio* **30,** 77-87.
DANSEREAU, P., BUELL, P. F. & DAGON, R. (1966). A universal system for recording vegetation. *Sarracenia* **10,** 1-64.
DIDAY, E. (1974). Optimization in non-heirarchical clustering. *Pattern Recognition* **6,** 17-33.
DORAN, J. (1968). New developments in the Graph Traverser. *In* Machine intelligence 2. Eds. E. Dale & D. Michie. Edinburgh: University Press, pp. 119-36.
EDWARDS, A. W. F. & CAVALLI-SFORZA, L. L. (1965). A method for cluster analysis. *Biometrics* **21,** 362-75.
EIGEN, D. J., FROMM, F. R. & NORTHOUSE, R. A. (1974). Cluster analysis based on dimensional information with applications to feature selection and classification. *IEEE Trans. Syst. Man. Cybern. Vol. SMC* **4,** pp. 284-94.
FURNIVAL, G. M. & WILSON Jr., R. W. (1974). Regressions by leaps and bounds. *Technometrics* **16,** 499-511.
GOODALL, D. W. (1953). Objective methods for the classification of vegetation. I. The use of positive interspecific correlation. *Aust. J. Bot.* **1,** 39-63.
GOODALL, D. W. (1954). Objective methods for the classification of vegetation. III. An essay in the use of factor analysis. *Aust. J. Bot.* **2,** 304-24.
GOODALL, D. W. (1969). A procedure for recognition of uncommon species combinations in sets of vegetation samples. *Vegetatio* **18,** 19-35.
GOWER, J. C. (1966). Some distance properties of latent root and vector methods used in multivariate analysis. *Biometrika* **53,** 325-38.
GRABAU, W. E. & RUSHING, W. N. (1968). A computer-compatible system for quantitatively describing the physiognomy of vegetation assemblages. *In* Land evaluation. Ed. G. A Stewart, Melbourne: Macmillan of Australia, pp. 263-75.
GREIG-SMITH, P. (1952). The use of random and contiguous quadrats in the study of the structure of plant communities. *Ann. Bot. NS.* **16,** 293-316.
GREIG-SMITH, P., AUSTIN, M. P. & WHITMORE, T. C. (1967). The application of quantitative methods to vegetation survey. I. Association-analysis and principal component ordination of rain forest. *J. Ecol.* **55,** 483-503.
HALL, P. A. V. (1971). Branch-and-bound and beyond. *Proc 2nd Int. Joint Conf. Artif. Intell.* pp. 641-50.
HARMAN, H. H. (1967). Modern factor analysis. Chicago: University Press.
HARRIS, L. R. (1974). The heuristic search under conditions of error. *Artif. Intell.* **5,** 217-34.
HARTIGAN, J. A. (1972). Direct clustering of a data matrix. *J. Amer. Stat. Assoc.* **67,** 123-129.
HILL, M. O. (1973) Reciprocal averaging: an eigenvector method of ordination. *J. Ecol.* **61,** 237-50.
HILL, M. O., BUNCE, R. C. H. & SHAW, M. W. (1975). Indicator species analysis, a divisive polythetic method of classification, and its application to a survey of native pinewoods in Scotland. *J. Ecol.* **63,** 597-615.
HOTELLING, H. (1933). Analysis of a complex of statistical variables into principal components. *J. Educ. Psych.* **24,** 417-41, 498-520.
IVIMEY-COOK, R. B. & PROCTOR, M. C. F. (1967). Factor analysis of data from an east

Devon heath: a comparison of principal component and rotated solutions. *J. Ecol.* **55,** 405-14.

JANCEY, R. C. (1974). Algorithm for detection of discontinuities in data sets. *Vegetatio* **29,** 131-3.

JARDINE, N. & SIBSON, R (1968). The construction of hierarchic and non-hierarchic classifications. *Comput. J.* **11,** 177-84.

JARDINE, N. & SIBSON, R. (1971). Mathematical taxonomy. London: John Wiley & Sons.

JENKINS, G. M. & WATTS, D. G. (1968). Spectral analysis and its applications. San Francisco: Holden-Day.

KERSHAW, K. A. (1960). The detection of pattern and association. *J. Ecol.* **48,** 233-42.

KOONTZ, W. L. G., NARENDRA, P. M. & FUKUNAGA, K. (1975). A branch and bound clustering algorithm. *IEEE Trans. Comput.* *C*-24, pp. 908-15.

KRUSKAL, J B. (1964). Nonmetric multidimensional scaling: a numerical method. *Psychometrika* **29,** 115-29.

LACOSTE, A. & ROUX, M. (1971) L'analyse multidimensionnelle en phytosociologie et en écologie. I. L'analyse des données floristiques. *Oecol. Plant.* **6,** 353-69.

LACOSTE, A. & ROUX, M. (1972). L'analyse multidimensionnelle en phytosociologie et en écologie. II. L'analyse des données écologiques et l'analyse globale. *Oecol. Plant* **7,** 125-46.

LAMBERT, J. M. (1972). Theoretical modes for large-scale vegetation survey. *In* Mathematical models in ecology. Ed. J. N. R. Jeffers, Oxford: Blackwell Scientific Publications pp. 87-109.

LAMBERT, J. M. & DALE, M. B. (1964). The use of statistics in phytosociology. *Adv. Ecol. Res.* **2,** 59-99.

LAMBERT, J. M., MEACOCK, S. E., BARRS, J. & SMARTT, P. F. M. (1973). AXOR and MONIT: Two new polythetic-divisive strategies for hierarchical classification. *Taxon.* **22,** 173-6.

LAMBERT, J. M. & WILLIAMS, W. T. (1962). Multivariate methods in plant ecology. IV. Nodal analysis. *J. Ecol.* **50,** 775-802.

LANCE, G. N. (1970). Mixed and discontinuous data. *In* Data representation. Eds. R. S. Anderssen & M. R. Osborne. Brisbane: University of Queensland Press, pp. 102-7.

LANCE, G. N. & WILLIAMS, W. T. (1966). A generalized sorting strategy for computer classifications. *Nature, UK,* **212,** 218.

LANCE, G. N. & WILLIAMS, W. T. (1967). Mixed-data classificatory programs. 1. Agglomerative systems. *Aust. Comput. J.* **1,** 15-20.

LANCE, G. N. & WILLIAMS, W. T. (1968). Note on a new information-statistic classificatory program. *Comput. J.* **11,** 195.

LANGE, R. T. (1968). Influence analysis in vegetation. *Aust. J. Bot.* **16,** 555-64.

MACKAY, D. M. (1969). Recognition and action. *In* Methodologies of pattern recognition. Ed. S. Watanabe. London: Academic Press, pp. 409-16.

MACNAUGHTON-SMITH, P. (1965). Some statistical and other numerical techniques for classifying individuals. *Home Office Res. Unit Rep. No.* 6, London: HMSO.

MACNAUGHTON-SMITH, P., WILLIAMS, W. T., DALE, M. B. & MOCKETT, L. G. (1964). Dissimilarity analysis: a new technique of hierarchical sub-division. *Nature, UK* **202,** 1034-5.

MCQUITTY, L. L. (1968). Multiple clusters, types, and dimensions from iterative intercolumnar correlational analysis. *Multivar. Behav. Res.* **3,** 465-77.

MANTEL, N. & VALAND, R. S. (1970). A technique of non-parametric multivariate analysis. *Biometrics* **26,** 547-58.

MORRIS, D. K. (1972). An introduction to wheel systems. Proc. Vth Aust. Comput. Conf., pp. 623-32.

NORRIS, J. M. & DALE, M. B. (1971). Transition matrix approach to numerical classification of soil profiles. *Soil Sci. Soc. Am. Proc.* **35,** 487-91.

NOTLEY, M. G. (1970). The cumulative recurrence library. *Comput. J.* **13,** 14-9.

NOY-MEIR, I. (1971). Multivariate analysis of the semi-arid vegetation in South-Eastern Australia: Nodal ordination by component analysis. *Proc. Ecol. Soc. Aust.* **6,** 159-93.

NOY-MEIR, I. (1973a). Data transformations in ecological ordination. I. Some advantages of non-centering. *J. Ecol.* **61,** 329-42.

NOY-MEIR, I. (1973b). Divisive polythetic classification of vegetation data by optimized division on ordination components. *J. Ecol.* **61,** 753-60.

NOY-MEIR, I. (1974). Multivariate analysis of the semiarid vegetation in South-eastern Australia. II. Vegetation catenae and environmental gradients. *Aust. J. Bot.* **22,** 115-40

NOY-MEIR, I. & ANDERSON, D. J (1971). Multiple pattern analysis, or multiscale ordination: towards a vegetation hologram ? *In* Statistical ecology, Vol. 3, Eds. G. P. Patil, E. C. Pielou & W. E. Waters. Pennsylvania: State University Press, pp. 207-31.

ORLÓCI, L. (1966). Geometric models in ecology. I. The theory and application of some ordination methods. J. *Ecol.* **54,** 193-215.

ORLÓCI, L. (1967). An agglomerative method for classification of plant communities. *J. Ecol.* **55,** 193-206.

ORLÓCI, L. (1970). Analysis of vegetation samples based on the use of information. *J. Theoret. Biol.* **29,** 173-89.

ORLÓCI, L. (1975). Multivariate analysis in vegetation research. The Hague: Dr. W. Junk, N.V.

POHL, I. (1969). First results on the effect of error in heuristic search. *In* Machine intelligence 5. Eds. B. Meltzer & D. Michie. Edinburgh: University Press, pp. 219-36.

POHL, I. (1970). Bi-directional search. *In* Machine intelligence 6. Eds. B Meltzer & D. Michie. Edinburgh: University Press, pp. 127-40.

SHEPARD, R. N. & CARROLL, J. D. (1966). Parametric representation of nonlinear data structures. *In* Multivariate analysis. Ed. P. R. Krishnaiah. New York: Academic Press, pp. 561-92.

SLAGLE, J. R. (1970). Heuristic search programs. *In* Theoretical approaches to non-numerical problem solving. Eds. R. Banerji & M. D. Mesarovic. Berlin: Springer Verlag, pp. 246-73.

SMARTT, P. F. M., MEACOCK, S. E. & LAMBERT, J M. (1974). Investigations into the properties of quantitative vegetational data. I. Pilot Study. *J. Ecol.* **62,** 735-59.

SNEATH, P. H. A. & SOKAL, R. R. (1973). Numerical taxonomy. San Francisco: W. M. Freeman & Co.

SONQUIST, J. A., BAKER, E. L. & MORGAN, J. N. (1973). Searching for Structure. Univ. Mich. Ann Arbor Off. Res. Adm.

STEPHENSON, W., WILLIAMS, W. T. & LANCE, G. N. (1970). The macrobenthos of Moreton Bay. *Ecol. Monogr.* **40,** 459-94.

THARU, J. & WILLIAMS, W. T. (1966). Concentration of entries in binary arrays. *Nature, UK* 210, 549.

UTTLEY, A. M. (1970). The INFORMON: A network for adaptive pattern recognition. *J. Theoret. Biol.* **27,** 31-67.

VINOD, H. D. (1969). Integer programming and the theory of grouping. *J. Am. Stat. Assoc.* **64,** 506-19.

WEBB, L. J., TRACEY, J. C., WILLIAMS, W. T. & LANCE, G. N. (1967). Studies in the numerical analysis of complex rain-forest communities. II. The problem of species-sampling. *J. Ecol.* **55,** 525-38.

WEBB, L. J., TRACEY, J. G., WILLIAMS, W. T. & LANCE, G. N. (1970). Studies in the numerical analysis of complex rain-forest communities. V. A comparison of the properties of floristic and physiognomic-structural data. *J. Ecol.* **58,** 203-32.

WEBB, L. J., TRACEY, J. G., WILLIAMS, W. T. & LANCE, G. N. (1971). Prediction of agricultural potential from intact forest vegetation. *J. Appl. Ecol.* **8,** 99-121.

WEBSTER, J. T., GUNST, R. F. & MASON, R. L. (1974). Latent root regression analysis. *Technometrics.* **16,** 513-22.

WEIR, A. D. (1970). Program DIVINFRE: A divisive classification on binary data, including re-allocation. *Aust. CSIRO Div. Land Use Res. Tech. Mem.* 70/7.

WHITTAKER, R. H. (1973). Direct gradient analysis: Techniques. *In* Handbook of vegetation science. V. Ordination and classification of communities. Ed. R. H. Whittaker. The Hague: Dr. W. Junk NV.

WILLIAMS, W. T. (1969). The problem of attribute-weighting in numerical classification. *Taxon.* **18,** 369-74.

WILLIAMS, W. T. (1972). Partition of information. *Aust. J. Bot.* **20,** 235-40.

WILLIAMS, W. T. (1973). Partition of information: the CENTPERC problem. *Aust. J. Bot.* **21,** 277-81.

WILLIAMS, W. T. (Ed.) (1976). Pattern analysis in agricultural research. Melbourne: CSIRO and Elsevier Scientific Publishing Co.

WILLIAMS, W. T., CLIFFORD, H. T. & LANCE, G. N. (1972). Group-size dependence: a rationale for choice between numerical classifications. *Comput. J.* **14,** 157-62.

WILLIAMS, W. T. & DALE, M. B. (1965). Fundamental problems in numerical taxonomy. *Adv. Bot. Res.* **2,** 35-68.

WILLIAMS, W. T., DALE, M. B. & MACNAUGHTON-SMITH, P. (1964). An objective method of weighting in similarity analysis. *Nature, UK* **201,** 426.

WILLIAMS, W. T. & EDYE, L. A. (1974). A new method for the analysis of three-dimensional data matrices in agricultural experimentation. *Aust. J. Agric. Res.* **25,** 803-12.

WILLIAMS, W. T. & GILLARD, P. (1971). Pattern analysis of a grazing experiment. *Aust. J. Agric. Res.* **22,** 245-60.

WILLIAMS, W. T., HAYDOCK, K. P., EDYE, L. A. & RITSON, J. B. (1971a). Analysis of a fertility trial with Droughtmaster cows. *Aust. J. Agric. Res.* **22,** 979-91.

WILLIAMS, W. T. & LAMBERT, J. M. (1959). Multivariate methods in plant ecology. 1. Association-analysis in plant communities. *J. Ecol.* **47,** 83-101.

WILLIAMS, W. T., LAMBERT, J. M. & LANCE, G. N. (1966). Multivariate methods in plant ecology. V. Similarity analyses and information-analysis. *J. Ecol.* **54,** 427-45.

WILLIAMS, W. T., LANCE, G. N., DALE, M. B. & CLIFFORD, H. T. (1971b). Controversy concerning the criteria for taxonometric strategies. *Comput. J.* **14,** 162-5.

WILLIAMS, W. T., LANCE, G. N., WEBB, L. J., TRACEY, J. G. & CONNELL, J. H. (1969a). Studies in the numerical analysis of complex rain-forest communities. IV. A method for the elucidation of small-scale pattern. *J. Ecol.* **57,** 635-54.

WILLIAMS, W. T., LANCE, G. N., WEBB, L. J., TRACEY, J. G. & DALE, M. B. (1969b). Studies in the numerical analysis of complex rain-forest communities. III. The analysis of successional data. *J. Ecol.* **57,** 515-35.

WISHART, D. (1969). Mode analysis: a generalization of nearest neighbour which reduces chaining effects. *In* Numerical taxonomy. Ed. A. J. Cole, London: Academic Press, pp. 282-311.

YARRANTON, G. A., BEASLEIGH, W J., MORRISON, R. G. & SHAFI, M. I. (1972). On the classification of phytosociological data into non-exclusive groups with a conjecture about determining the optimal number of groups in a classification. *Vegetatio* **24,** 1-12.

ZAHN, C. T. (1971). Graph-theoretical methods for detecting and describing gestalt clusters. *IEEE Trans. Comput.* **C-20,** 68-86.

ZINNES, J. L. & GRIGGS, R. A. (1974). Probabilistic, multidimensional unfolding analysis. *Psychometrika* **39,** 327-50.

INDEX

CONVERSION TABLE OF WEIGHTS AND MEASURES

Length
1 metre (m) = 39·37 inches
1 kilometere (km) = 0·62 miles

Area
1 hectare (ha) = 2·47 acres

Weight
1 kilogram (kg) = 2·20 lb.
1 tonne (t) = 1000 kg

Volume
1 litre (l) = 0·220 Imp. gallon = 0·264 US gallon
1 kg/ha = 0·89 lb/acre